A COURSE OF BIOLOGY ENGLISH

钱国英 司爱侠 朱秋华 编著

生物专业英语教程

清华大学出版社
北京

内 容 简 介

本书以"Unit"为单位,每一"Unit"由以下几部分组成:"Text A"(课文 A)及"Text B"(课文 B)——这些课文语言地道,知识面广;"New Words"(单词)——对课文中出现的新词进行注释,读者由此可以积累生物专业的基本词汇;"Phrases"(词组)——对课文中的常用词组进行注释;"Abbreviations"(缩略语)——对课文中出现的、业内人士必须掌握的缩略语进行注释;"Notes"(难句讲解)——讲解课文中出现的疑难句子,培养读者的阅读理解能力;"Exercises"(习题)——供读者练习使用,可有效地巩固学习效果;"Reading Material"(阅读材料)——选编生物学专业文献,进一步扩大读者视野;"练习答案"——给出全书练习题答案,可供读者检查学习效果。本书以生物学常用知识为主线,运用概念解释、理论阐述、知识介绍等多种形式,帮助读者掌握生物学专业英语的基本术语和表达方式,能切实提高生物工作者实际运用专业英语的能力。

本书既可作为高等院校生物学及相关专业学生的专业英语教材,也可供其他生物学工作者"充电"之用。

版权所有,侵权必究。举报:010-62782989,beiqinquan@tup.tsinghua.edu.cn

图书在版编目(CIP)数据

生物专业英语教程/钱国英,司爱侠,朱秋华编著.—北京:清华大学出版社,2006.10(2024.8重印)
ISBN 978-7-302-13595-1

Ⅰ.生… Ⅱ.①钱… ②司… ③朱… Ⅲ.生物学-英语-高等学校-教材 Ⅳ.H31

中国版本图书馆 CIP 数据核字(2006)第 090978 号

责任编辑:罗 健
责任印制:刘海龙

出版发行:清华大学出版社
网　　址:https://www.tup.com.cn, https://www.wqxuetang.com
地　　址:北京清华大学学研大厦 A 座　　邮　编:100084
社 总 机:010-83470000　　邮　购:010-62786544
投稿与读者服务:010-62776969,c-service@tup.tsinghua.edu.cn
质 量 反 馈:010-62772015,zhiliang@tup.tsinghua.edu.cn

印 装 者:三河市龙大印装有限公司
经 销 者:全国新华书店

开　本:185mm×260mm　　印　张:13.25　　字　数:312 千字
版　次:2006 年 10 月第 1 版　　印　次:2024 年 8 月第 17 次印刷
定　价:39.80 元

产品编号:017631-03/Q

前　　言

随着我国与外界交流日益增加,对生物专业从业人员专业英语水平的要求也日益提高,因此,生物专业从业人员必须进行针对性的专门学习。编著者编写本书的目的就在于切实提高读者实际使用生物专业英语的能力。

本书体例上以"Unit"为单位,每一"Unit"由以下几部分组成:"Text A"(课文 A)及"Text B"(课文 B)——这些课文语言地道,知识面广;"New Words"(单词)——对课文中出现的新词进行注释,读者由此可以积累生物专业的基本词汇;"Phrases"(词组)——对课文中的常用词组进行注释;"Abbreviations"(缩略语)——对课文中出现的、业内人士必须掌握的缩略语进行注释;"Notes"(难句讲解)——讲解课文中出现的疑难句子,培养读者的阅读理解能力;"Exercises"(习题)——供读者练习使用,可有效巩固学习效果;"Reading Material"(阅读材料)——选编生物学专业文献,进一步扩大读者视野;"练习答案"——给出全书练习题答案,可供读者检查学习效果。

本书内容比较全面,涉及生物学多个分支学科领域。以生物学常用基本知识为主线,运用概念解释、理论阐述、知识介绍等多种形式,帮助读者掌握生物学专业英语的基本术语和表达方式,能切实提高生物工作者实际运用专业英语的能力。

本书结构非常适合组织教学,词汇加注了音标。

本书既可作为高等院校生物学及相关专业学生的专业英语教材,也可供其他生物学工作者"充电"之用。

作　者
2006 年 8 月

目录

Unit 1
- [1] Text A   Biology: The Science of Our Lives
- [2]   New Words
- [5]   Phrases
- [6]   Abbreviations
- [6]   Notes
- [7]   Exercises
- [9] Text B   Characteristics of Living Things
- [12]   New Words
- [14]   Phrases
- [15]   Abbreviations
- [15]   Exercises
- [16] Reading Material   Primates and Their Adaptations
- [21] Text A 参考译文   生物学：生命的科学
- [21] Text B 参考译文   生物的特性

Unit 2
- [24] Text A   Mitosis
- [26]   New Words
- [27]   Phrases
- [27]   Notes
- [28]   Exercises
- [31] Text B   Cells
- [33]   New Words
- [35]   Phrases
- [36]   Exercises
- [36] Reading Material   The Nucleus and Cytoplasm
- [39] Text A 参考译文   有丝分裂
- [41] Text B 参考译文   细胞

Unit 3
- [43] Text A   The Modern View of the Genes
- [45]   New Words

[46]	Phrases	
[47]	Notes	
[47]	Exercises	
[50]	Text B	The Human Karyotype
[53]	New Words	
[54]	Phrases	
[54]	Exercises	
[55]	Reading Material	Polygenic Inheritance and Chromosomes
[57]	Text A 参考译文	基因的现代观点
[58]	Text B 参考译文	人类的染色体组型

Unit 4

[61]	Text A	Microevolution and Macroevolution
[62]	New Words	
[64]	Phrases	
[65]	Notes	
[65]	Exercises	
[68]	Text B	Darwinian Evolution
[69]	New Words	
[71]	Phrases	
[71]	Exercises	
[72]	Reading Material	Darwin's Critics
[74]	Text A 参考译文	小进化和大进化
[75]	Text B 参考译文	达尔文进化论

Unit 5

[77]	Text A	Plants and Their Structure
[79]	New Words	
[80]	Phrases	
[80]	Notes	
[81]	Exercises	
[83]	Text B	Flowers
[86]	New Words	
[88]	Phrases	
[88]	Exercises	
[89]	Reading Material	Plant Balanced Diet
[92]	Text A 参考译文	植物与植物结构
[93]	Text B 参考译文	花

Unit 6

| [96] | Text A | Organization of the Animal Body |

[98]   New Words
[100] Phrases
[100] Abbreviations
[100] Notes
[101] Exercises
[103] Text B   Some General Features of Animals
[105] New Words
[106] Phrases
[106] Exercises
[107] Reading Material Animal Organ Systems and Homeostasis
[112] Text A 参考译文   动物体组织
[114] Text B 参考译文   动物的一般特征

Unit 7

[116] Text A   Ecosystems
[118] New Words
[119] Phrases
[120] Abbreviations
[120] Notes
[120] Exercises
[123] Text B   Terrestrial Biomes
[126] New Words
[127] Phrases
[128] Exercises
[128] Reading Material Community and Ecosystem Dynamics
[131] Text A 参考译文   生态系统
[133] Text B 参考译文   陆地生物群落

Unit 8

[135] Text A   Amino Acid Analysis
[136] New Words
[137] Phrases
[137] Notes
[138] Exercises
[141] Text B   Glycoprotein Biosynthesis and Function
[143] New Words
[144] Phrases
[144] Exercises
[145] Reading Material Hydrogen Bonding
[147] Text A 参考译文   氨基酸分析

V

[148] Text B 参考译文　　糖蛋白生物合成和功能

Unit 9
[151]　Text A　　　　　　Immune Response
[153]　New Words
[154]　Phrases
[154]　Abbreviations
[155]　Notes
[155]　Exercises
[158]　Text B　　　　　　Cellular Counterattack: The Second Line of Defense
[160]　New Words
[160]　Phrases
[161]　Exercises
[162]　Reading Material　The Immune Response: The Third Line of Defense
[164]　Text A 参考译文　　免疫反应
[166]　Text B 参考译文　　细胞反击：第二道防线

Unit 10
[168]　Text A　　　　　　Stem Cells
[169]　New Words
[170]　Phrases
[170]　Notes
[171]　Exercises
[174]　Text B　　　　　　Gene Therapy
[175]　New Words
[176]　Phrases
[176]　Exercises
[177]　Reading Material　Should We Label Genetically Modified Foods
[179]　Text A 参考译文　　干细胞
[180]　Text B 参考译文　　基因治疗

练习答案
[181]　Unit 1
[183]　Unit 2
[185]　Unit 3
[187]　Unit 4
[189]　Unit 5
[191]　Unit 6
[193]　Unit 7
[195]　Unit 8
[197]　Unit 9
[199]　Unit 10

Unit 1

Text A

Biology: The Science of Our Lives

Biology literally means "the study of life". Biology is such a broad field, covering the minute workings of chemical machines inside our cells, to broad scale concepts of ecosystems and global climate change. Biologists study intimate details of the human brain, the composition of our genes, and even the functioning of our reproductive system. Biologists recently all but completed the deciphering of the human genome, the sequence of deoxyribonucleic acid (DNA) bases that may determine much of our innate capabilities and predispositions to certain forms of behavior and illnesses. DNA sequences have played major roles in criminal cases (O. J. Simpson, as well as the reversal of death penalties for many wrongfully convicted individuals), as well as the impeachment of President Clinton (the stain at least did not lie). We are bombarded with headlines about possible health risks from favorite foods (Chinese, Mexican, hamburgers, etc.) as well as the potential benefits of eating other foods such as cooked tomatoes. Informercials tout the benefits of metabolism-adjusting drugs for weight loss. Many Americans are turning to herbal remedies to ease arthritis pain, improve memory, as well as improve our moods.

Robert Hooke (1635-1703), one of the first scientists to use a microscope to examine pond water, cork and other things, referred to the cavities he saw in cork as "cells", Latin for chambers. Mattias Schleiden (in 1838) concluded all plant tissues consisted of cells. In 1839, Theodore Schwann came to a similar conclusion for animal tissues. Rudolf Virchow, in 1858, combined the two ideas and added that all cells come from pre-existing cells, formulating the Cell Theory. Thus there is a chain-of-existence extending from your cells back to the earliest cells, over 3.5 billion years ago. The cell theory states that all organisms are composed of one or more cells, and that those cells have arisen from pre-existing cells.

In 1953, James Watson and Francis Crick developed the model for deoxyribonucleic acid (DNA), a chemical that had (then) recently been deduced to be the physical carrier of inheritance. Crick hypothesized the mechanism for DNA replication and further linked DNA to proteins, an idea since referred to as the central dogma. Information from DNA "language" is converted into RNA (ribonucleic acid) "language" and then to the "language" of proteins. The central dogma explains the influence of heredity (DNA) on the organism (proteins).

Homeostasis is the maintenance of a dynamic range of conditions within which the organism can function. Temperature, pH and energy are major components of this concept. Thermodynamics is a field of study that covers the laws governing energy transfers, and thus the basis for life on earth. Two major laws are known: the conservation of matter and energy, and entropy. The universe is composed of two things: matter (atoms, etc.) and energy.

These first three theories are very accepted by scientists and the general public. The theory of evolution is well accepted by scientists and most of the general public. However, it remains a lightening rod for school boards, politicians, and television preachers. Much of this confusion results from what the theory says and what it does not say.

New Words

biology	[baiˈɔlədʒi]	n. 生物学,生物
science	[ˈsaiəns]	n. 科学;学科
literally	[ˈlitərəli]	adv. 照字面意义,逐字地
broad	[brɔːd]	adj. 宽的,阔的,广泛的
field	[fiːld]	n. 领域
minute	[maiˈnjuːt]	adj. 微小的,异常小的,极细微的
working	[ˈwəːkiŋ]	n. 工作,工作方式
		adj. 工作的
chemical	[ˈkemikəl]	adj. 化学的
		n. 化学制品,化学药品
machine	[məˈʃiːn]	n. 机器,机械,组织
cell	[sel]	n. 细胞;单元
scale	[skeil]	n. 范围
concept	[ˈkɔnsept]	n. 观念,概念
ecosystem	[iːkəˈsistəm]	n. 生态系统
global	[ˈgləubəl]	adj. 全球的,全世界的
climate	[ˈklaimit]	n. 气候;风土,思潮
intimate	[ˈintimit]	adj. 亲密的;个人的,私人的
detail	[ˈdiːteil]	n. 细节,详情

2

Unit 1

composition	[kɔmpə'ziʃən]	n.	组成,构成
gene	[dʒi:n]	n.	遗传因子,遗传基因
functioning	['fʌŋkʃəniŋ]	n.	机能
reproductive	['ri:prə'dʌktiv]	adj.	生殖的,再生的
biologist	[bai'ɔlədʒist]	n.	生物学家
decipher	[di'saifə]	vt.	译解(密码等);解释
genome	['dʒi:nəum]	n.	基因组,染色体组
sequence	['si:kwəns]	n.	次序,顺序,序列
determine	[di'tə:min]	vt.	决定,确定,测定
innate	['ineit]	adj.	先天的,天生的
capability	[ˌkeipə'biliti]	n.	能力,性能
predisposition	[pri:ˌdispə'ziʃən]	n.	倾向;素质;诱因
behavior	[bi'heivjə]	n.	举止,行为
illness	['ilnis]	n.	疾病,生病
reversal	[ri'və:səl]	n.	颠倒,反转,反向,逆转
wrongfully	['rɔŋfuli]	adv.	错误地,不正当地,不妥地
convict	['kɔnvikt]	vt.	证明……有罪,宣告……有罪
individual	[ˌindi'vidjuəl]	n.	个人,个体
		adj.	个别的,单独的,个人的
impeachment	[im'pi:tʃmənt]	n.	非难;指责
headline	['hedlain]	n.	大字标题
risk	[risk]	n.	冒险,风险
		vt.	冒……的危险
favorite	['feivərit]	adj.	喜爱的,宠爱的,中意的
		n.	特别喜欢的人,喜欢的事物,亲信,心腹,幸运儿
hamburger	['hæmbə:gə]	n.	汉堡包;牛肉饼
potential	[pə'tenʃ(ə)l]	adj.	潜在的,可能的
		n.	潜能,潜力
benefit	['benifit]	n.	利益,好处
		vt.	有益于,有助于
		vi.	受益
informercial	[ˌinfə'mɜ:ʃəl]	n.	商业信息片
tout	[taut]	vt.	吹嘘,吹捧促销或激情洋溢地赞扬
drug	[drʌg]	n.	药,药物,药材
herbal	['hə:bəl]	adj.	草药的
remedy	['remidi]	n.	药物,治疗法,补救
		vt.	治疗,补救

英文	音标	释义
ease	[i:z]	vt. 使悠闲,使安心,减轻
		n. 安逸,安心,不费力,悠闲
arthritis	[ɑːˈθraitis]	n. 关节炎
improve	[imˈpruːv]	vt. 改善,改进
memory	[ˈmeməri]	n. 记忆,记忆力
mood	[muːd]	n. 心情,情绪
microscope	[ˈmaikrəskəup]	n. 显微镜
examine	[igˈzæmin]	vt. 研究,分析
pond	[pɔnd]	n. 池塘
cork	[kɔːk]	n. 软木塞,软木
cavity	[ˈkæviti]	n. 洞,孔;空穴
Latin	[ˈlætin]	n. 拉丁文,拉丁语
		adj. 拉丁文的,拉丁语的,拉丁人的
chamber	[ˈtʃeimbə]	n. 腔,室;房间
formulate	[ˈfɔːmjuleit]	vt. 对……作简洁陈述,有系统地表达;把……作为公式,用公式表示
state	[steit]	vt. 声明,陈述,规定
		n. 状况;情形;状态
organism	[ˈɔːgənizəm]	n. 生物体,有机体
deduce	[diˈdjuːs]	vt. 推论,演绎出
physical	[ˈfizikəl]	adj. 物质的,自然的,物理的
carrier	[ˈkæriə]	n. 携带者,带菌者;带虫者;传染疾病的媒介
inheritance	[inˈheritəns]	n. 遗传,继承;遗产
hypothesize	[haiˈpɔθisaiz]	vt. 假设,假定,猜测
		vi. 假定,作一个假设
mechanism	[ˈmekənizəm]	n. 结构;机械装置;机件
replication	[ˌrepliˈkeiʃən]	n. 复制
protein	[ˈprəutiːn]	n. 蛋白质
		adj. 蛋白质的
heredity	[hiˈrediti]	n. 遗传;遗传性;遗传特征
homeostasis	[ˌhəumiəuˈsteisis]	n. 体内平衡;自我平衡;动态平衡;内环境稳定
maintenance	[ˈmenˈteinəns]	n. 维持,保持,维护
dynamic	[daiˈnæmik]	adj. 动态的;动力的;有活力的;精力充沛的;有生命力的
temperature	[ˈtempritʃə(r)]	n. 温度;体温
pH	[piːeitʃ]	n. (表示氢离子活度的)pH 值

energy	['enədʒi]	n.	精力,精神,活力,能量
component	[kəm'pəunənt]	n.	成分,部分
		adj.	组成的,构成的
thermodynamics	['θəːməudai'næmiks]	n.	热力学
entropy	['entrəpi]	n.	熵;宇宙在能量与物质平均扩散后的状态,(热、能源的)扩散及消失
universe	['juːnivəːs]	n.	宇宙,世界,万物
matter	['mætə]	n.	物质;物体;事情,问题
politician	[pɔli'tiʃən]	n.	政治家,政客
preacher	['priːtʃə]	n.	讲道者;传教士;说教者

Phrases

reproductive system	再生系统,繁殖系统
all but	几乎;差一点
criminal case	刑事案件
as well as	也
death penalty	死刑
play roles in...	在……中起作用
be bombarded with	连珠炮似地,连珠炮般地
weight loss	减肥,重量减轻
turn to	转向
refer to... as	把……看作
plant tissue	植物组织,植物性组织
consist of...	由……组成
come to a conclusion	得出结论
animal tissue	动物组织
pre-existing cell	预有的细胞
cell theory	细胞理论
be composed of...	由……组成
arise from	由……而引起,由……而产生;从……中产生
central dogma	中心法则
convert... into	把……转换成
energy transfer	能量转移
the conservation of matter and energy	物质能量守恒定律
the theory of evolution	进化论
school board	学校董事会

result from... 由……而造成

Abbreviations

DNA (deoxyribonucleic acid)　　　　脱氧核糖核酸
RNA (ribonucleic acid)　　　　　　　核糖核酸

Notes

[1] Biologists recently all but completed the deciphering of the human genome, the sequence of deoxyribonucleic acid (DNA) bases that may determine much of our innate capabilities and predispositions to certain forms of behavior and illnesses.

本句中, the sequence of deoxyribonucleic acid (DNA) bases that may determine much of our innate capabilities and predispositions to certain forms of behavior and illnesses 是一个名词短语, 做 the human genome 的同位语, 对其作进一步补充说明。在该名词短语中, that may determine much of our innate capabilities and predispositions to certain forms of behavior and illnesses 是一个定语从句, 修饰和限定 the sequence of deoxyribonucleic acid (DNA) bases。

[2] Robert Hooke (1635-1703), one of the first scientists to use a microscope to examine pond water, cork and other things, referred to the cavities he saw in cork as "cells", Latin for chambers.

本句中, one of the first scientists to use a microscope to examine pond water, cork and other things 是 Robert Hooke 的同位语, 说明其身份。he saw in cork 是一个定语从句, 修饰和限定 the cavities, Latin for chambers 是 cells 的同位语, 对其进行解释。

[3] In 1953, James Watson and Francis Crick developed the model for deoxyribonucleic acid (DNA), a chemical that had (then) recently been deduced to be the physical carrier of inheritance.

本句中, a chemical that had (then) recently been deduced to be the physical carrier of inheritance 是一个名词短语, 做 deoxyribonucleic acid (DNA) 的同位语, 对其作进一步补充说明。在该名词短语中, that had (then) recently been deduced to be the physical carrier of inheritance 是一个定语从句, 修饰和限定 a chemical。

[4] Homeostasis is the maintenance of a dynamic range of conditions within which the organism can function.

本句中, within which the organism can function 是一个介词前置的定语从句, 修饰和限定 conditions。

[5] Thermodynamics is a field of study that covers the laws governing energy transfers,

and thus the basis for life on earth.

本句中，that covers the laws governing energy transfers, and thus the basis for life on earth 是一个定语从句，修饰和限定 a field of study。在该定语从句中，governing energy transfers 是一个现在分词短语，作定语，修饰和限定 the laws。

Exercises

【EX. 1】 根据课文内容，回答以下问题

1) What does biology literally mean?

2) What does biology cover?

3) What do biologists study?

4) What is deoxyribonucleic acid (DNA)?

5) What does the cell theory state?

6) Who developed the model for deoxyribonucleic acid and when?

7) What is homeostasis?

8) What are the major components of homeostasis?

9) What is thermodynamics?

10) what are the two laws mentioned in the passage?

【EX. 2】 根据下面的英文解释，写出相应的英文词汇（使用本单元所学的单词、词组或缩略语）

英　文　解　释	词　汇
The science of life and of living organisms, including their structure, function, growth, origin, evolution, and distribution. It includes botany and zoology and all their subdivisions.	

续表

英 文 解 释	词 汇
The smallest structural unit of an organism that is capable of independent functioning, consisting of one or more nuclei, cytoplasm, and various organelles, all surrounded by a semipermeable cell membrane.	
The state of being predisposed; tendency, inclination, or susceptibility.	
A commercial television program or relatively long commercial segment offering consumer information, such as educational or instructional material, related to the sponsor's product or service.	
The act or process by which genetic material, a cell, or an organism reproduces or makes an exact copy of itself.	
The genetic transmission of characteristics from parent to offspring.	
An individual that carries one gene for a particular recessive trait.	
The ability or tendency of an organism or a cell to maintain internal equilibrium by adjusting its physiological processes.	
An ecological community together with its environment, functioning as a unit.	
To reach (a conclusion) by reasoning.	

【EX. 3】 把下列句子翻译为中文

1) The core principle of biology is that biological diversity is the result of a long evolutionary journey.

2) Science is a way of viewing the world that focuses on objective information, putting that information to work to build understanding.

3) All living things share certain key characteristics: order, sensitivity, growth, development and reproduction, regulation, and homeostasis.

4) Living things are highly organized, whether as single cells or as multicellular organisms, with several hierarchical levels.

5) Both scientists and lay people are drawn to biology, because it seeks to answer the question of how life began.

6) The more related two species of multicellular organisms are, the more similar their anatomies in almost all cases.

7) Much emphasis in biology is in biotechnology, the use of organisms to create products.

8) At the same time, these prospects will challenge scientists with serious ethical considerations in the years to come, as the use of biotechnology requires scientists to manipulate the course of evolution.

9) Virtually every organism uses the same genetic code to builds its proteins, from the tiniest bacterium to the blue whale and the giant sequoia.

10) Biologists use a variety of technical and conceptual tools to study living things.

【EX. 4】 把下列短文翻译为中文

Biology is a fascinating and important subject, because it dramatically affects our daily lives and our futures. Many biologists are working on problems that critically affect our lives, such as the world's rapidly expanding population and diseases like cancer and AIDS. The knowledge these biologists gain will be fundamental to our ability to manage the world's resources in a suitable manner, to prevent or cure diseases, and to improve the quality of our lives and those of our children and grandchildren.

Biology is one of the most successful of the "natural sciences", explaining what our world is like. To understand biology, you must first understand the nature of science. The basic tool a scientist uses is thought. To understand the nature of science, it is useful to focus for a moment on how scientists think. They reason in two ways: deductively and inductively.

Text B

Characteristics of Living Things

Living things have a variety of common characteristics.

Organization. Living things exhibit a high level of organization, with multicellular organisms being subdivided into cells, and cells into organelles, and organelles into molecules, etc.

Homeostasis. Homeostasis is the maintenance of a constant yet also dynamic internal environment in terms of temperature, pH, water concentrations, etc. Much of our own metabolic energy goes toward keeping within our own homeostatic limits. If you run a high fever for long enough, the increased temperature will damage certain organs and impair your proper functioning. Swallowing of common household chemicals, many of which are outside the pH (acid/base) levels we can tolerate, will likewise negatively impact the human body's homeostatic regime. Muscular activity generates heat as a waste product. This heat is removed from our bodies by sweating. Some of this heat is used by warm-blooded animals, mammals and birds, to maintain their internal temperatures.

Adaptation. Living things are suited to their mode of existence. Charles Darwin began the recognition of the marvelous adaptations all life has that allow those organisms to exist in their environment.

Reproduction and heredity. Since all cells come from existing cells, they must have some way of reproducing, whether that involves asexual (no recombination of genetic material) or sexual (recombination of genetic material). Most living things use the chemical DNA (deoxyribonucleic acid) as the physical carrier of inheritance and the genetic information. Some organisms, such as retroviruses, of which HIV is a member, use RNA (ribonucleic acid) as the carrier. The variation that Darwin and Wallace recognized as the wellspring of evolution and adaptation is greatly increased by sexual reproduction.

Growth and development. Even single-celled organisms grow. When first formed by cell division, they are small, and must grow and develop into mature cells. Multicellular organisms pass through a more complicated process of differentiation and organogenesis because they have so many more cells to develop.

Energy acquisition and release. One view of life is that it is a struggle to acquire energy from sunlight, inorganic chemicals, or another organism, and release it in the process of forming ATP, that is adenosine triphosphate.

Detection and response to stimuli both internal and external.

Interactions. Living things interact with their environment as well as each other. Organisms obtain raw materials and energy from the environment or another organism. The various types of symbioses, that is the organismal interactions with each other, are examples of this.

There are four classes of macromolecules in living things. They are polysaccharides, triglycerides, polypeptides and nucleic acids. These classes perform a variety of functions in cells.

Carbohydrates have the general formula $[CH_2O]_n$ where n is a number between 3 and 6. Carbohydrates function as short-term energy storage (such as sugar); as intermediate-term energy storage (starch for plants and glycogen for animals); and as structural components in cells (cellulose in the cell walls of plants and many protists, and chitin in the exoskeleton of insects and other arthropods).

Sugars are structurally the simplest carbohydrates. They are the structural units which makes up the other types of carbohydrates. Monosaccharides are single sugars. Important monosaccharides include ribose ($C_5H_{10}O_5$), glucose ($C_6H_{12}O_6$), and fructose which has the same formula as glucose but different in structure.

Lipids are involved mainly with long-term energy storage. They are generally insoluble in polar substances such as water. Secondary functions of lipids are as structural components, as in the case of phospholipids that are the major building block in cell membranes and as "messengers", such as hormones, that play roles in communications within and between cells. Lipids are composed of three fatty acids usually covalently bonded to a 3-carbon glycerol. The fatty acids are composed of CH_2 units, and are hydrophobic, not water-soluble.

Proteins are very important in biological systems as control and structural elements. Control functions of proteins are carried out by enzymes and proteinaceous hormones. Enzymes are chemicals that act as organic catalysts. A catalyst is a chemical that promotes but is not changed by a chemical reaction. Structural proteins function in the cell membrane, muscle tissue, etc.

The building block of any protein is the amino acid, which has an amino end (NH_2) and a carboxyl end (COOH). The R indicates the variable component (R-group) of each amino acid. Alanine and valine, for example, are both nonpolar amino acids, but they differ, as do all amino acids, by the composition of their R-groups. All living things and even viruses use various combinations of the same twenty amino acids.

Nucleic acids are polymers composed of monomer units known as nucleotides. There are a very few different types of nucleotides. The main functions of nucleotides are information storage (DNA), protein synthesis (RNA), and energy transfers (ATP and NAD). Nucleotides consist of a sugar, a nitrogenous base, and a phosphate. The sugars are either ribose or deoxyribose. They differ by the lack of one oxygen in deoxyribose. Both are pentoses usually in a ring form. There are five nitrogenous bases. Purines (adenine and guanine) are double-ring structures, while pyrimidines (cytosine, thymine and uracil) are single-ringed.

New Words

characteristic	[ˌkærɪktə'rɪstɪk]	n. 特性,特征
		adj. 特有的,表示特性的,典型的
exhibit	[ɪg'zɪbɪt]	vt. 展示;展出,陈列
		n. 展览品,陈列品,展品
organization	[ˌɔːgənaɪ'zeɪʃən]	n. 组织,机构,团体,构成
multicellular	[ˌmʌltɪ'seljulə]	adj. 多细胞的
subdivide	['sʌbdɪ'vaɪd]	vt. 再分,细分
organelle	[ˌɔːgə'nel]	n. 细胞器
molecule	['mɔlɪkjuːl]	n. 分子
concentration	[ˌkɔnsen'treɪʃən]	n. 含量,浓度;浓缩
metabolic	[ˌmetə'bɔlɪk]	adj. 代谢作用的,新陈代谢的
organ	['ɔːgən]	n. 器官
impair	[ɪm'pɛə]	vt. 削弱,使弱;损害
functioning	['fʌŋkʃənɪŋ]	n. 机能
swallow	['swɔləu]	vt. 吞下,咽下
tolerate	['tɔləreɪt]	vt. 忍受,容忍
impact	['ɪmpækt]	vt. 撞击;冲击
		n. 冲击(力);碰撞;影响
muscular	['mʌskjulə]	adj. 肌肉的,强健的
generate	['dʒenəˌreɪt]	vt. 产生,发生
sweat	[swet]	vt. 使出汗
		vi. 出汗
		n. 汗,汗珠
recognition	[ˌrekəg'nɪʃən]	n. 承认,认可,公认
marvelous	['mɑːvɪləs]	adj. 绝妙的;了不起的
asexual	[æ'seksjuəl]	adj. 无性的,无性生殖的
sexual	['seksjuəl]	adj. 性的,性别的,有性的
recombination	['riːkɔmbɪ'neɪʃən]	n. 再结合,重组
retrovirus	[ˌretrəu'vaɪərəs]	n. [微]逆转录酶病毒(一种致肿瘤病毒)
wellspring	['welˌsprɪŋ]	n. 水源,源泉
differentiation	[ˌdɪfəˌrenʃɪ'eɪʃən]	n. 分化,变异;区别,差别
release	[rɪ'liːs]	n. 释放
		vt. 释放
organogenesis	[ˌɔːgənəu'dʒenɪsɪs]	n. 器官发生,器官形成

Unit 1

acquire	[ə'kwaiə]	vt.	获得,取得;学到
detection	[di'tekʃən]	n.	察觉,发觉,发现
response	[ris'pɔns]	n.	回答,响应,反应
stimulus	['stimjuləs]	n.	刺激物,促进因素,刺激 stimuli(pl)
interaction	[ˌintə'rækʃən]	n.	交互作用;互相影响
symbiosis	[simbai'əusis]	n.	共生(现象),合作(或互利,互依)关系,symbioses(pl)
macromolecule	[ˌmækrəu'mɔlikju:l]	n.	大分子,高分子
carbohydrate	[kɑ:bəu'haidreit]	n.	糖;碳水化合物
glycogen	['glaikəudʒen]	n.	糖原
cellulose	['seljuləus]	n.	纤维素;植物纤维质
protist	['prəutist]	n.	原生生物
chitin	['kaitin]	n.	几丁质,壳多糖,壳素
exoskeleton	[ˌeksəu'skelitən]	n.	外骨骼
insect	['insekt]	n.	昆虫
arthropod	['ɑ:θrəpɔd]	n.	节肢动物
		adj.	节肢动物的
structurally	['strʌktʃərəli]	adv.	结构上
monosaccharide	[ˌmɔnəu'sækəraid]	n.	单糖
ribose	['raibəus]	n.	核糖
glucose	['glu:kəus]	n.	葡萄糖
fructose	['frʌktəus]	n.	果糖
lipid	['lipid]	n.	脂类;脂质,油脂
insoluble	[in'sɔljubl]	adj.	不能溶解的
phospholipid	[ˌfɔsfəu'lipid]	n.	磷脂
hormone	['hɔ:məun]	n.	荷尔蒙,激素
communication	[kəmju:ni'keiʃ(ə)n]	n.	通信;联络;传送
covalently	[kəu'veiləntli]	adv.	共价地
glycerol	['glisərɔl]	n.	甘油,丙三醇
soluble	['sɔljubl]	adj.	可溶的,可溶解的
enzyme	['enzaim]	n.	酶
proteinaceous	[ˌprəuti:'neiʃəs]	adj.	蛋白质的,似蛋白质的
organic	[ɔ:'gænik]	adj.	器官的,有机的,组织的
catalyst	['kætəlist]	n.	催化剂
promote	[prə'məut]	vt.	促进
amino	['æminəu]	adj.	氨基的
carboxyl	[kɑ:'bɔksil]	n.	羧基

alanine	[ˈæləˌniːn]	n. 丙胺酸
valine	[ˈvæliːn]	n. 缬氨酸
nonpolar	[ˈnɔnˈpəulə]	adj. 无极的,非极性的
phylogenetic	[ˌfailəudʒəˈnetik]	adj. 系统发生的,动植物种类史的
polymer	[ˈpɔlimə]	n. 聚合体
monomer	[ˈmɔnəmə]	n. 单体
nucleotide	[ˈnjuːkliətaid]	n. 核苷
phosphate	[ˈfɔsfeit]	n. 磷酸盐
deoxyribose	[diːɔksiˈraibəus]	n. 脱氧核糖
pentose	[ˈpentəus]	n. 戊糖
purine	[ˈpjuəriːn]	n. 嘌呤
adenine	[ˈædəniːn]	n. 腺嘌呤
guanine	[ˈgwɑːniːn]	n. 鸟嘌呤
pyrimidine	[ˌpaiəˈrimidiːn]	n. 嘧啶
cytosine	[ˈsaitəsiːn]	n. 胞嘧啶
thymine	[ˈθaimiːn]	n. 胸腺嘧啶
uracil	[ˈjuərəsil]	n. 尿嘧啶

Phrases

a variety of	许多
multicellular organism	多细胞有机体
in terms of	根据,依照
sexual reproduction	有性生殖,有性繁殖
cell division	细胞分裂
mature cell	成熟细胞
pass through	穿过
energy acquisition	能量获得
inorganic chemical	无机化学成分
interact with…	与……相合,与……互相作用
energy storage	能量储存
cell wall	细胞壁
polar substance	极性物质
building block	骨架
cell membrane	细胞膜
fatty acid	脂肪酸
chemical reaction	化学反应

structural protein	结构蛋白
muscle tissue	肌肉组织
amino acid	氨基酸
amino end	氨基端
carboxyl end	羧基端
nucleic acid	核酸
information storage	信息储存
protein synthesis	蛋白合成
energy transfer	能量转换
a nitrogenous base	含氮碱基
in a ring form	呈环形地

Abbreviations

HIV (human immunodeficiency virus)	人体免疫缺陷病毒,艾滋病病毒
ATP (Adenosine Triphosphate)	三磷酸腺苷
NAD (nicotinamide adenine dinucleotide)	烟酰胺腺嘌呤二核苷酸,辅酶Ⅰ

Exercises

【EX. 5】 根据课文内容,回答以下问题

1) What is homeostasis?

2) What do most living things use the chemical DNA (deoxyribonucleic acid) as?

3) Why do ulticellular organisms pass through a more complicated process of differentiation and organogenesis?

4) How many classes of macromolecules are there in living things? What are they?

5) What are sugars?

6) What is a catalyst?

7) What is the building block of any protein?

8) What are nucleic acids?

9) What are the main functions of nucleotides?

10) What does nucleotides consist of?

Reading Material

Text	Notes
Primates[1] and Their Adaptations 　　Mammals[2] developed from primitive mammal-like reptiles[3] during the Triassic Period[4], some 200-245 million years ago. After the terminal Cretaceous[5] extinction (65 million years ago) eliminated the dinosaurs[6], mammals as one of the surviving groups, underwent an adaptive radiation[7] during the Tertiary Period. The major orders of mammals developed at this time, including the Primates to which humans belong. 　　Other primates include the tarsiers[8], lemurs[9], gibbons[10], monkeys, and apes. Although we have significant differences from other primates, we share an evolutionary history that includes traits such as opposable thumbs, stereoscopic vision, larger brains, and nails replacing claws. 　　Primates are relatively unspecialized mammals: they have no wings, still have all four limbs, cannot run very fast, have generally weak teeth, and lack armor or thick protective hides[11]. However, the combination of primate adaptations that include larger brains, tool use, social structure, stereoscopic color vision, highly developed forelimbs and hands, versatile teeth, and upright posture, places them among the most advanced mammals (at least as judged from an anthrocentric perspective!). 　　Approximately 20 million years ago, central and east Africa was densely forested. Climatic changes resulting from plate tectonic[12] movements and episodes of global cooling about 15 million years ago caused a replacement of the forest by a drier-adapted savanna[13] mixed with open areas of forest. During the course	[1]灵长类 [2]哺乳动物 [3]爬虫动物 [4]三叠纪 [5]白垩纪的 [6]恐龙 [7]适应性辐射 [8]眼镜猴,跗猴 [9]狐猴 [10]长臂猿 [11]兽皮,动物的皮 [12]板块构造 [13]热带大草原

Text	Notes
of hominid evolution, periodic climate changes would trigger bursts of evolution and/or extinction.	
Primates have modifications to their ulna[14] and radius[15] (bones of the lower arm) allowing them to turn their hand without needing to turn their elbow. Many primates can also swivel or turn their arms at the shoulder. These two adaptations offer advantages to life in the trees.	[14]尺骨 [15]桡骨
Primates have five digits[16] on their forelimbs. They are able to grasp objects with their forelimbs in what is known as a prehensile movement. A second modification makes one of the digits opposable, allowing the tips of the fingers and thumb to touch.	[16]手指或足趾
Placement of the eyes on the front of the head increases depth perception, an advantageous trait in tree-dwelling primates. Changes in the location of rods and cones in the eye adapted primates for color vision as well as peripheral vision in dim light.	
Upright posture allows a primate to view its surroundings as well as to use its hands for some other task. Hominids, the lineage leading to humans, had changes in the shape and size of their pelvis[17], femur[18], and knees that allowed bipedalism[19] (walking on two legs). The change from quadruped to biped happened in stages, culminating in humans, who can walk or run on two legs.	[17]骨盆 [18]大腿骨,腿节 [19]双足行走(运动)
Several trends of primate evolution are evident in the teeth and jaw. First, change in the geometry of the jaw reduced the snout into a flat face. Second, changes in tooth arrangement and numbers increased the efficiency of those teeth for grinding food. Third, about 1.5 million years ago our diet changed from fruits and vegetables to include meat.	
Origin of Apes and Hominids The fossil record indicates primates evolved about approximately 30 million years ago in Africa. One branch of primates evolved into the Old and New World Monkeys, the	

Text	Notes
other into the hominoids[20] (the line of descent common to both apes and man). Fossil hominoids occur in Africa during the Miocene Epoch[21] of the Tertiary Period[22]. They gave rise to an array of species in response to major climate fluxes in their habitats. However, the nature of those habitats leads to an obscuration of the line that leads to humans (the hominids). Until a few years ago, the ramapiths[23] were thought to have given rise to the hominids. We now consider ramapiths ancestral to the orang-outang. The hominid line arose from some as-yet-unknown ancestor. Lacking fossil evidence, biochemical and DNA evidence suggests a split of the hominid from hominoid line about 6 to 8 million years ago. *Australopithecus*[24] *afarensis*, the first of the human-like hominids we know of, first appeared about 3.6-4 million years ago. This species had a combination of human (bipedalism) and apelike features (short legs and relatively long arms). The arm bones were curved like chimps, but the elbows were more human-like. Scientists speculate that *A. afarensis* spent some time climbing trees, as well as on the ground. *Australopithecus ramidus* is an older species, about 4.4 million years, and is generally considered more anatomically primitive than *A. afarensis*. The relationship between the two species remains to be solved. The role of *A. afarensis* as the stem from which the other hominids arose is in some dispute. About 2 million years ago, after a long million-year period of little change, as many as six hominid species evolved in response to climate changes associated with the beginning of the Ice Age. Two groups developed: the australopithecines, generally smaller brained and not users of tools; and the line that led to genus Homo, larger brained and makers and users of tools. The australopithecines died out 1 million years ago; Homo, despite their best efforts (atomic weapons, pollution)	[20]类人动物 [21]中新世 [22]第三纪 [23]腊玛古猿 [24]南猿,南方古猿

18

Text	Notes
are still here!	
With an incomplete fossil record, australopithecines, at least the smaller form, *A. africanus*, was thought ancestral to Homo. Recent discoveries, however, have caused a reevaluation of that hypothesis. One pattern is sure, human traits evolved at different rates and at different times, in a mosaic[25]: some features (skeletal, dietary) establishing themselves quickly, others developing later (tool-making, language, use of fire).	[25]嵌合体
A cluster of species developed about 2-2.5 million years ago in Africa. *Homo* had a larger brain and a differently shaped skull and teeth than the australopithecines. About 1.8 million years ago, early *Homo* gave rise to *Homo erectus*, the species thought to have been ancestral to our own.	
Soon after its origin (1.8 million but probably older than 2 million years ago) in Africa, Homo erectus appears to have migrated out of Africa and into Europe and Asia. Homo erectus differed from early species of Homo in having a larger brain size, flatter face, and prominent brow ridges. Homo erectus is similar to modern humans in size, but has some differences in the shape of the skull, a receding chin, brow ridges, and differences in teeth. Homo erectus was the first hominid to:	
1. provide evidence for the social and cultural aspects of human evolution.	
2. leave Africa (living in Africa, Europe, and Asia).	
3. use fire.	
4. have social structures for food gathering.	
5. utilize permanent settlements.	
6. provide a prolonged period of growth and maturation[26] after birth.	[26]成熟
Between 100,000 and 500,000 years ago, the world population of an estimated 1 million *Homo erectus* disappeared, replaced by a new species, *Homo sapiens*[27]. How, when and where this new species arose and how it replaced	[27]人类

Text	Notes

its predecessor remain in doubt. Answering those questions has become a multidisciplinary task.
 	
  Two hypotheses differ on how and where Homo sapiens originated.
 	
  1. The Out-of-Africa Hypothesis proposes that some *H. erectus* remained in Africa and continued to evolve into *H. sapiens*, and left Africa about 100,000-200,000 years ago. From a single source, *H. sapiens* replaced all populations of H. erectus. Human populations today are thus all descended from a single speciation event in Africa and should display a high degree of genetic similarity. Support for this hypothesis comes from DNA studies of mitochondria[28]: since African populations display the greatest diversity of mitochondrial DNA, modern humans have been in Africa longer than they have been elsewhere. Calculations suggest all modern humans are descended from a population of African H. sapiens numbering as few as 10,000.
 	
  2. The Regional Continuity Hypothesis suggests that regional populations of *H. erectus* evolved into *H. sapiens* through interbreeding[29] between the various populations. Evidence from the fossil record and genetic studies supports this idea.
 	
  Which hypothesis is correct? Scientists can often use the same "evidence" to support contrasting hypotheses depending on which evidence (fossils or molecular clock/DNA studies) one gives more weight to. The accuracy of the molecular clock, so key to the out-of-Africa hypothesis, has recently been questioned. Recent studies on the Y-chromosome seem to weaken the regional continuity hypothesis by indicating a single point-of-origin for our species some 270,000 years ago. Continued study will no doubt reveal new evidence and undoubtedly new hypotheses will arise. It is a task for all of us to weigh the evidence critically and reach a supportable conclusion, whether we are scientists or not.
	[28]线粒体
	
	[29]杂交繁殖

Text A 参考译文

生物学：生命的科学

从字面上理解，生物学的意思是研究生命的科学。生物学的研究领域是如此的宽广，涵盖的范围从人类细胞内极细微活动的化学机制，到生态系统的广泛概念和全球气候变化。生物学家研究人类大脑的奥秘、基因的组成，甚至还有生殖系统的机能。最近，生物学家快要完成人类基因组编码信息，即脱氧核糖核酸(DNA)碱基序列的测定工作，这些碱基序列也许基本上决定了我们的先天能力以及一些行为和疾病的倾向性。DNA 序列分析已经在刑事案件侦破中起主要作用(例如 O. J. 辛普森案件，以及许多误判罪犯的死刑撤销方面的案件)，在克林顿总统的弹劾事件中也起了主要作用(血迹不会说谎)。一些宣传向我们扑面而来，例如吃风味食品(中餐、墨西哥食品、汉堡包等)可能会给我们带来健康隐患，而吃其他食品，比如煮熟的西红柿，会有潜在的益处。一些商业信息片面夸大调节新陈代谢的药物对减肥的效果。许多美国人转而采用草药疗法来减轻关节炎的疼痛、提高记忆力、调节情绪。

罗伯特·虎克(1635—1703)是最早使用显微镜来检测池塘水、软木和其他东西的科学家之一，他把在软木中所见到的腔称为"细胞"，在拉丁语中"细胞"表示小室的意思。Mattias Schleiden(1838) 认为所有的植物组织都由细胞组成。1839 年，Theodore Schwann 认为动物组织也由细胞组成。1858 年，Rudolf Virchow 把这两种观点结合起来，他认为所有的细胞都来自于先前存在的细胞，这就是细胞理论。由这一理论可知，从你的细胞中可以追溯到最早的细胞，联结它们的"链"跨度超过 35 亿年。细胞理论认为所有的有机体都是由一个或多个细胞组成，且这些细胞都源于已存在的细胞。

在 1953 年，詹姆士·沃森和弗朗西斯·克里克发现了脱氧核糖核酸(DNA)的双螺旋结构模型。后来，DNA 被推导出是一种遗传信息实质载体的化学物质。克里克提出了 DNA 复制与从 DNA 到蛋白质的假说机制，这一想法后来被称为中心法则。DNA 转录信息到 RNA，RNA 翻译为蛋白质。中心法则解释了遗传物质(DNA)对有机体(蛋白质)的影响。

动态平衡是指有机体可以发挥功能时的动态条件的保持。这些动态条件主要是指温度、pH 和能量。热力学主要研究能量从一种形式转换为另一种形式时遵从的规律，这也是地球上生命的基础。已知的主要规律有两条：物质和能量守恒定律及熵的守恒定律。宇宙万物都是由两样东西组成的，即物质(原子等)和能量。

上述前三条理论已经被科学家和公众真正接受。进化理论也基本上被科学家和大部分公众所接受。但是对学术界、政客和电视传媒而言，它仍然是争论的焦点，主要原因是难以区分这个理论所表达的东西和没有表达的东西是什么。

Text B 参考译文

生物的特性

生物有多种多样的共性。下文将一一列举：

机体组织结构。生物展示了其高级的机体组织结构，多细胞的有机体被分解为细胞，细胞再分解为细胞器，细胞器可被再分为分子等。

内环境平衡。内环境平衡是指恒定而又动态的内部环境平衡的维持,其因素包括温度、pH、含水量等。我们大部分的代谢能都被用于自身体内环境平衡的维持。如果你发高烧太久,升高的体温会破坏特定器官以及削弱你的正常机能。许多普通家用化学药品的pH(酸/碱)值超出我们所能承受的水平,吞服这些药品同样也会对人体的内环境平衡产生不良的影响。肌肉活动产生的热作为一种人体消耗产物,这种热通过出汗的方式被排出体外。一些恒温动物(哺乳动物和鸟类)可利用产生的部分热来维持体温。

适应性。生物与它们的生存方式相适应。查尔斯·达尔文首先认识到所有生命都具有奇异的适应性,这就使得这些有机体可在它们所处的环境中生存。

繁殖和遗传。由于所有的细胞都源于已存在的细胞,那么它们就一定有繁殖的方法(无性繁殖(无遗传物质的重组),或有性繁殖(有遗传物质的重组))。大多数生物把化学物质DNA(脱氧核糖核酸)作为遗传特征以及遗传信息的物质载体。一些生物体,比如逆转录酶病毒(HIV就属于这种病毒)把RNA(核糖核酸)作为遗传物质的载体。达尔文和华莱士认识到生物的变异性是进化和适应的源泉,这种变异性通过有性繁殖得到了极大的增强。

生长和发育。即使是单细胞生物生长,当通过细胞分裂开始形成时,细胞很小,必须要通过生长和发育来形成成熟的细胞。多细胞有机体因为有更多的细胞需要发育,要经过一个更加复杂的分化和器官发生过程。

能量获得和释放。关于生命的一个理论是尽力获得能量(从太阳光、无机化合物或其他有机物中),再在形成ATP(三磷酸腺苷)的过程中释放出来。

内外刺激的察觉和反应。

交互作用。生物与其环境相互作用,生物与生物之间也相互作用。生物体从环境或其他生物体处获取原材料和能量。各种类型的共生就是生物之间相互作用的例子。

生物体有4类大分子,即多糖、三酰甘油、多肽和核酸。这些不同类型的大分子在细胞中起着不同的作用。

碳水化合物的通式是$[CH_2O]_n$,这里n的取值在3至6之间。碳水化合物有能量短期储存功能(如糖)和能量中期储存功能(植物中的淀粉和动物中的糖原);也作为细胞的结构成分(植物和许多原生生物细胞壁中的纤维素,昆虫和其他节肢动物外骨骼中的壳多糖)。

单糖是结构最简单的碳水化合物,它们是组成其他类型碳水化合物的结构单元。单糖只含一个糖基,重要的单糖包括核糖($C_5H_{10}O_5$)、葡萄糖($C_6H_{12}O_6$)和果糖(与葡萄糖比较,化学式相同但是结构不同)。

脂肪与能量长期储存有关。它们通常不溶于极性物质(比如水)。脂类的另一个功能是作为结构组分,如磷脂是细胞壁的主要结构单元,并在细胞内和细胞之间作为"信使物质"(如激素),起到细胞内及细胞之间通信的作用。脂肪通常由三种脂肪酸共价结合到甘油上而组成。脂肪酸是由CH_2单元组成,具疏水性,不溶于水。

蛋白质作为调控物质和结构物质,对生物系统非常重要。蛋白质的调控功能由酶和蛋白质类激素来完成。酶是一类具有机催化剂作用的化学物质。催化剂是一种化学物质,这种化学物质可促进化学反应,但不会被化学反应改变。结构蛋白质在细胞壁、肌肉组织等中起作用。

所有蛋白质的结构单元都是氨基酸,氨基酸具有一个氨基端(NH_2)和一个羧基端(COOH)。R 表示每一氨基酸的可变成分(R 基团),例如丙氨酸和缬氨酸都是非极性氨基酸,但它们的区别在于 R 基团组成的不同,这也是所有的氨基酸得以区别的地方。所有的生物(甚至病毒)都是由进行不同的组合的同样的 20 种的氨基酸组成。

核酸是由已知为核苷酸的单体组成的聚合体。核苷酸的种类很少,其主要作用是信息储藏(DNA)、蛋白质合成(RNA)和能量转移(ATP 和 NAD)。核苷酸由一个糖、一个含氮碱基和一个磷酸组成。糖可以是核糖,也可以是脱氧核糖,两者的区别是脱氧核糖少了一个氧原子。核糖和脱氧核糖都是戊糖,且通常呈环状。含氮碱基有 5 种类型,嘌呤(腺嘌呤和鸟嘌呤)是双环结构,而嘧啶(胞嘧啶、胸腺嘧啶和尿嘧啶)是单环结构。

Unit 2

Text A

Mitosis

1. Prophase: Formation of the Mitotic Apparatus

When the chromosome condensation initiated in G2 phase reaches the point at which individual condensed chromosomes first become visible with the light microscope, the first stage of mitosis, prophase, has begun. The condensation process continues throughout prophase; consequently, some chromosomes that start prophase as minute threads appear quite bulky before its conclusion. Ribosomal RNA synthesis ceases when the portion of the chromosome bearing the rRNA genes is condensed.

Assembling the Spindle Apparatus. The assembly of the microtubular apparatus that will later separate the sister chromatids also continues during prophase. In animal cells, the two centriole pairs formed during G2 phase begin to move apart early in prophase, forming between them an axis of microtubules referred to as spindle fibers. By the time the centrioles reach the opposite poles of the cell, they have established a bridge of microtubules called the spindle apparatus between them. In plant cells, a similar bridge of microtubular fibers forms between opposite poles of the cell, although centrioles are absent in plant cells.

During the formation of the spindle apparatus, the nuclear envelope breaks down and the endoplasmic reticulum reabsorbs its components. At this point, then, the microtubular spindle fibers extend completely across the cell, from one pole to the other. Their orientation determines the plane in which the cell will subsequently divide, through the center of the cell at right angles to the spindle apparatus.

In animal cell mitosis, the centrioles extend a radial array of microtubules toward the plasma membrane when they reach the poles of the cell. This arrangement of microtubules is called an aster. Although the aster's function is not fully understood, it probably braces the centrioles against the membrane and stiffens the point of microtubular attachment during the retraction of the spindle. Plant cells, which have rigid cell walls, do not form asters.

Linking Sister Chromatids to Opposite Poles. Each chromosome possesses two kinetochores, one attached to the centromere region of each sister chromatid. As prophase continues, a second group of microtubules appears to grow from the poles of the cell toward the centromeres. These microtubules connect the kinetochores on each pair of sister chromatids to the two poles of the spindle. Because microtubules extending from the two poles attach to opposite sides of the centromere, they attach one sister chromatid to one pole and the other sister chromatid to the other pole. This arrangement is absolutely critical to the process of mitosis; any mistakes in microtubule positioning can be disastrous. The attachment of the two sides of a centromere to the same pole, for example, leads to a failure of the sister chromatids to separate, so that they end up in the same daughter cell.

2. Metaphase: Alignment of the Centromeres

The second stage of mitosis, metaphase, is the phase where the chromosomes align in the center of the cell. When viewed with a light microscope, the chromosomes appear to array themselves in a circle along the inner circumference of the cell, as the equator girdles the earth. An imaginary plane perpendicular to the axis of the spindle that passes through this circle is called the metaphase plate. The metaphase plate is not an actual structure, but rather an indication of the future axis of cell division. Positioned by the microtubules attached to the kinetochores of their centromeres, all of the chromosomes line up on the metaphase plate. At this point, which marks the end of metaphase, their centromeres are neatly arrayed in a circle, equidistant from the two poles of the cell, with microtubules extending back towards the opposite poles of the cell in an arrangement called a spindle because of its shape.

3. Anaphase and Telophase: Separation of the Chromatids and Reformation of the Nuclei

Of all the stages of mitosis, anaphase is the shortest and the most beautiful to watch. It starts when the centromeres divide. Each centromere splits in two, freeing the two sister chromatids from each other. The centromeres of all the chromosomes separate simultaneously, but the mechanism that achieves this synchrony is not known. Freed from each other, the sister chromatids are pulled rapidly toward the poles to which their kinetochores are attached. In the process, two forms of movement take place simultaneously, each driven by microtubules. First, the poles move apart as microtubular spindle fibers physically anchored to opposite poles slide past each other, away from the center of the cell. Because another group of microtubules attach the chromosomes to the poles, the chromosomes move apart, too. If a flexible membrane surrounds the cell, it becomes visibly elongated. Second, the centromeres move toward the poles as the microtubules that connect them to the poles shorten. This shortening process is not a contraction; the microtubules do not get any thicker. Instead, tubulin subunits are removed from the kinetochore ends of the microtubules by the organizing center. As more subunits are removed, the chromatid-bearing microtubules are progressively disassembled, and the chromatids are pulled ever closer to the poles of the cell.

When the sister chromatids separate in anaphase, the accurate partitioning of the replicated genome—the essential element of mitosis—is complete. In telophase, the spindle apparatus disassembles, as the microtubules are broken down into tubulin monomers that can be used to construct the cytoskeletons of the daughter cells. A nuclear envelope forms around each set of sister chromatids, which can now be called chromosomes because each has its own centromere. The chromosomes soon begin to uncoil into the more extended form that permits gene expression. One of the early group of genes expressed are the rRNA genes, resulting in the reappearance of the nucleolus.

New Words

mitosis	[mi'təusis]	n.	有丝分裂
prophase	['prəufeiz]	n.	前期,初期,早期
chromosome	['krəuməsəum]	n.	染色体
condensation	[ˌkɔnden'seiʃən]	n.	浓缩
initiate	[i'niʃieit]	v.	开始
bulky	['bʌlki]	adj.	大的
ribosomal	[ˌraibə'səuməl]	adj.	核糖体的
synthesis	['sinθisis]	n.	合成
microtubule	[ˌmaikrəu'tju:bju:l]	n.	微管
orientation	[ˌɔ(:)rien'teiʃən]	n.	方向
aster	['æstə]	n.	星状体
centriole	['sentriəul]	n.	细胞的中心粒,中心粒
brace	[breis]	vt.	支持,使固定
stiffen	['stifn]	vt.	使硬
retraction	[ri'trækʃən]	n.	缩回;取消
kinetochore	[ki'ni:təkɔ:]	n.	动粒,着丝粒,着丝点
centromere	['sentrəmiə]	n.	着丝点,着丝粒
metaphase	['metəfeiz]	n.	中期
circumference	[sə'kʌmfərəns]	n.	周围
equator	[i'kweitə]	n.	赤道
girdle	['gə:dl]	vt.	束;绕,围绕;包围
		n.	带,腰带

equidistant	[ˌiːkwiˈdistənt]	adj. 等距离的
anaphase	[ˈænəˌfeiz]	n. 后期
simultaneously	[siməlˈteiniəsly]	adv. 同时地
mechanism	[ˈmekənizəm]	n. 机制
synchrony	[ˈsiŋkrəni]	n. 同步性,同步
anchor	[ˈæŋkə]	vi. 紧固;紧紧扣牢
elongated	[ˈiːlɔŋgeitid]	adj. 变长的
progressively	[prəˈgresivli]	adv. 逐渐的
disassemble	[ˌdisəˈsembl]	vt. 散开,解开,分解
uncoil	[ˈʌnˈkɔil]	v. 展开,解开
reappearance	[riəˈpiərəns]	n. 再现

Phrases

mitotic apparatus	有丝分裂装置
light microscope	光学显微镜
minute thread	微丝
spindle apparatus	纺锤体装置
microtubular apparatus	微管装置
sister chromatid	姐妹染色单体
centriole pair	配对中心粒
spindle fiber	纺锤纤维,纺锤体丝
nuclear envelope	核膜
endoplasmic reticulum	内质网
plasma membrane	细胞膜,质膜
daughter cell	子细胞
be perpendicular to	与……垂直
metaphase plate	中期板
tubulin subunit	微管蛋白亚基
remove... from...	把……从……中移去
gene expression	基因表达
result in	导致,终于造成……结果

Notes

[1] When the chromosome condensation initiated in G2 phase reaches the point at which individual condensed chromosomes first become visible with the light microscope, the

first stage of mitosis, prophase, has begun.

本句中,When the chromosome condensation initiated in G2 phase reaches the point at which individual condensed chromosomes first become visible with the light microscope 是一个时间状语从句。在该从句中,initiated in G2 phase 是一个过去分词短语,作定语,修饰和限定 the chromosome condensation; at which individual condensed chromosomes first become visible with the light microscope 是一个定语从句,修饰和限定 the point。

[2] In animal cells, the two centriole pairs formed during G2 phase begin to move apart early in prophase, forming between them an axis of microtubules referred to as spindle fibers.

本句中,formed during G2 phase 是一个过去分词短语,作定语,修饰和限定 the two centriole pairs。forming between them an axis of microtubules referred to as spindle fibers 是一个现在分词短语,作伴随状语。在该状语中,referred to as spindle fibers 是一个过去分词短语,作定语,修饰和限定 an axis of microtubules。

[3] At this point, which marks the end of metaphase, their centromeres are neatly arrayed in a circle, equidistant from the two poles of the cell, with microtubules extending back towards the opposite poles of the cell in an arrangement called a spindle because of its shape.

本句中,which marks the end of metaphase 是一个非限定性定语从句,对 at this point 进行补充说明。equidistant from the two poles of the cell 和 with microtubules extending back towards the opposite poles of the cell in an arrangement called a spindle because of its shape 分别是形容词短语和 with 的复合结构,对主句 their centromeres are neatly arrayed in a circle 起补充说明。

在 with 的复合结构中,called a spindle 是一个过去分词短语,作定语,修饰和限定 an arrangement,because of its shape 作原因状语。

[4] In telophase, the spindle apparatus disassembles, as the microtubules are broken down into tubulin monomers that can be used to construct the cytoskeletons of the daughter cells.

本句中,as the microtubules are broken down into tubulin monomers that can be used to construct the cytoskeletons of the daughter cells 是一个原因状语从句。在该从句中,that can be used to construct the cytoskeletons of the daughter cells 是一个定语从句,修饰和限定 tubulin monomers。

Exercises

【EX.1】 根据课文内容,回答以下问题
1) When has the first stage of mitosis, prophase, begun?

2) When does ribosomal RNA synthesis cease?

3) How many kinetochores does each chromosome possess?

4) Why do microtubules attach one sister chromatid to one pole and the other sister chromatid to the other pole?

5) Is microtubule positioning very critical? Give an example.

6) Is the metaphase plate an actual structure? What is it?

7) Of all the stages of mitosis, which is the shortest and the most beautiful to watch?

8) Why do the chromosomes move apart?

9) When is the accurate partitioning of the replicated genome complete?

10) What are one of the early group of genes expressed?

【EX. 2】 根据下面的英文解释,写出相应的英文词汇

英 文 解 释	词 汇
The process in cell division by which the nucleus divides.	
A threadlike linear strand of DNA and associated proteins in the nucleus of animal and plant cells that carries the genes and functions in the transmission of hereditary information.	
One of two cylindrical cellular structures that are composed of nine triplet microtubules and form the asters during mitosis.	
The stage of mitosis and meiosis in which the chromosomes move to opposite ends of the nuclear spindle.	
Any of the proteinaceous cylindrical hollow structures that are distributed throughout the cytoplasm of eukaryotic cells, providing structural support and assisting in cellular locomotion and transport.	

续表

英文解释	词汇
The first stage of mitosis, during which the chromosomes condense and become visible, the nuclear membrane breaks down, and the spindle apparatus forms at opposite poles of the cell.	
Set going by taking the first step; begin.	
The most condensed and constricted region of a chromosome to which the spindle fiber is attached during mitosis.	
A minute, round particle composed of RNA and protein found in the cytoplasm of living cells and active in the synthesis of proteins.	
Made longer; extended.	

【EX. 3】 把下列句子翻译为中文

1) All cells come from preexisting cells and have certain processes, types of molecules, and structures in common.

2) The first cells may have arisen from aggregates of macromolecules in bubbles.

3) To maintain adequate exchanges with its environment, a cell's surface area must be large compared with its volume.

4) All prokaryotic cells have a plasma membrane, a nucleoid region with DNA, and a cytoplasm that contains ribosomes, water, and dissolved proteins and small molecules.

5) Like prokaryotic cells, eukaryotic cells have a plasma membrane, cytoplasm, and ribosomes. However, eukaryotic cells are larger and contain many membrane-enclosed organelles.

6) The nucleus contains most of the cell's DNA, which associates with protein to form chromatin.

7) Multicellular organisms usually consist of many small cells rather than a few large ones because small cells function more efficiently. They have a greater relative surface area, enabling more rapid communication between the center of the cell and the environment.

8) Bacteria are small cells that lack interior organization. They are encased by an exterior wall composed of carbohydrates cross-linked by short polypeptides, and some are propelled by rotating flagella.

9) Multicellular diploid and multicellular haploid individuals both develop from single-celled beginnings by mitotic divisions.

10) Ribosomes are the sites of protein synthesis in the cytoplasm.

【EX. 4】 把下列短文翻译为中文

Organisms with complete extra sets of chromosomes may sometimes be produced by artificial breeding or by natural accidents. Under some circumstances, triploid ($3n$), tetraploid ($4n$), and higher-order polyploid nuclei may form. Each of these ploidy levels represents an increase in the number of complete sets of chromosomes present. If a nucleus has one or more extra full sets of chromosomes, its abnormally high ploidy in itself does not prevent mitosis. In mitosis, each chromosome behaves independently of the others. In meiosis, by contrast, homologous chromosomes must synapse to begin division. If even one chromosome has no homolog, anaphase I cannot send representatives of that chromosome to both poles. A diploid nucleus can undergo normal meiosis; a haploid one cannot. Similarly, a tetraploid nucleus has an even number of each kind of chromosome, so each chromosome can pair with its homolog. But a triploid nucleus cannot undergo normal meiosis, because one-third of the chromosomes would lack partners.

Text B

Cells

A. Classification of cells and organisms

Cells are the basic units of living organisms, with the exception of viruses whose structure and function are different from cells. Although a nerve cell looks entirely different from a red blood cell, their organizations are essentially the same. Even plant cells and animal cells share significant similarity in the overall organization.

All cells are divided into two types: prokaryotic cells and eukaryotic cells. The prokaryotic cell does not have a nucleus. The eukaryotic cell contains a nucleus.

Eukaryotes are the organisms made up of eukaryotic cells. They include protista, fungi, animals and plants. Prokaryotes include archaebacteria and eubacteria. They are single-cell organisms.

More recently, "archaebacteria" have been placed in a category outside "bacteria", because they are quite different from the ordinary bacteria. According to the new classification, prokaryotes are divided into archaea and bacteria, where "archaea" is equivalent to "archaebacteria", and "bacteria" is the same as "eubacteria".

Archaea live in extreme environments. They may be organized into three groups:

Methanogens live in anaerobic environment such as swamps. They produce methane and cannot tolerate exposure to oxygen.

Extreme halophiles live in very high concentrations of salt (NaCl), e. g., the Dead Sea and the Great Salt Lake.

Extreme thermophiles live in hot, sulfur rich and low pH environment, such as hot springs, geysers and fumaroles in the Yellowstone National Park.

B. Basic cellular components

All cells contain cytoplasm, plasma membrane, and DNA.

Cytoplasm is the viscous contents of a cell, including proteins, ribosomes, metabolites and ions. Ribosomes are the sites of protein synthesis.

Plasma membrane is the cell membrane surrounding cytoplasm. It consists of phospholipid bilayer, associated proteins and carbohydrates. Phospholipid bilayer is also the basic constituent of other biomembranes.

DNA (deoxyribonucleic acid) is the genetic material. An eukaryotic cell contains several DNA molecules, located in the nucleus and mitochondria which are membrane-bound organelles. A prokaryotic cell contains a single DNA molecule, which has no specific boundary with the cytoplasm.

C. The Cell Membrane and Cell Wall

The cell membrane functions as a semi-permeable barrier, allowing a very few molecules across it while fencing the majority of organically produced chemicals inside the cell. Electron microscopic examinations of cell membranes have led to the development of the lipid bilayer model (also referred to as the fluid-mosaic model). The most common molecule in the model is the phospholipid, which has a polar (hydrophilic) head and two nonpolar (hydrophobic) tails. These phospholipids are aligned tail to tail so the nonpolar areas form a hydrophobic region between the hydrophilic heads on the inner and outer surfaces of the membrane. This layering is termed a bilayer since an electron microscopic technique known as freeze-fracturing is able to split the bilayer.

Cholesterol is another important component of cell membranes embedded in the hydrophobic areas of the inner (tail-tail) region. Most bacterial cell membranes do not contain cholesterol.

Proteins are suspended in the inner layer, although the more hydrophilic areas of these proteins "stick out" into the cells interior and outside of the cell. These proteins function as gateways that will, in exchange for a price, allow certain molecules to cross into and out of the cell. These integral proteins are sometimes known as gateway proteins. The outer surface of the membrane will tend to be rich in glycolipids, which have their hydrophobic tails embedded in the hydrophobic region of the membrane and their heads exposed outside the cell. These, along with carbohydrates attached to the integral proteins, are thought to function in the recognition of self.

The contents (both chemical and organelles) of the cell are termed protoplasm, and are further subdivided into cytoplasm (all of the protoplasm except the contents of the nucleus) and nucleoplasm (all of the material, plasma and DNA etc. within the nucleus).

Not all living things have cell walls, most notably animals and many of the more animal-like protistans. Bacteria have cell walls containing peptidoglycan. Plant cells have a variety of chemicals incorporated in their cell walls. Cellulose is the most common chemical in the plant primary cell wall. Some plant cells also have lignin and other chemicals embedded in their secondary walls. The cell wall is located outside the plasma membrane. Plasmodesmata are connections through which cells communicate chemically with each other through their thick walls. Fungi and many protists have cell walls although they do not contain cellulose, rather a variety of chemicals (chitin for fungi).

New Words

similarity	[ˌsimiˈlæriti]	n. 类似,类似处
overall	[ˈəuvərɔːl]	adv. 大体上;一般地
		adj. 全部的,全面的
prokaryotic	[ˌprəuˌkæriˈɔtik]	adj. 原核的
eukaryotic	[ˌjuːkæriˈɔtik]	adj. 真核状态的
eukaryote	[juˈkæriəut]	n. 真核细胞
prokaryote	[prəuˈkæriəut]	n. 原核生物
archaebacteria	[ˌɑːkibækˈtiəriə]	n. 原始细菌,古细菌
eubacteria	[juːbækˈtiəriə]	n. 真细菌
archaea	[ɑːki]	n. 古细菌
bacteria	[bækˈtiəriə]	n. 细菌
methanogen	[ˈmeθənəˌdʒen]	n. 甲烷微生物
anaerobic	[ˌæneəˈrəubik]	adj. 厌氧性的
swamp	[swɔmp]	n. 沼泽;湿地;森林沼泽

methane	['meθein]	n.	甲烷,沼气
tolerate	['tɔləreit]	vt.	忍受,容忍
exposure	[iks'pəuʒə]	n.	暴露
halophile	['hæləfail]	n.	喜盐生物,适盐植物,喜盐植物
thermophile	['θə:məfail]	adj.	喜温的,嗜热的,适温的
		n.	嗜热生物,喜温生物
sulfur	['sʌlfə]	n.	硫磺;硫
geyser	['gaizə]	n.	天然喷泉;间歇泉
fumarole	['fju:mərəul]	n.	(火山地带的)喷气坑,气孔,喷气孔
cytoplasm	['saitəuplæzəm]	n.	细胞质
viscous	['viskəs]	adj.	黏性的,黏滞的,胶黏的
ribosome	['raibəsəum]	n.	核糖体
metabolite	[mi'tæbəlait]	n.	代谢物
ion	['aiən]	n.	离子
carbohydrate	['kɑ:bəu'haidreit]	n.	碳水化合物,糖类
constituent	[kən'stitjuənt]	n.	要素,组成部分
		adj.	形成的;组成的
biomembrane	[ˌbaiəu'membrein]	n.	生物膜
molecule	['mɔlikju:l]	n.	分子
nucleus	['nju:kliəs]	n.	核子[nuclear 的复数]
mitochondria	[ˌmaitə'kɔndriə]	n.	线粒体
membrane	['membrein]	n.	膜,隔膜
boundary	['baundəri]	n.	边界,分界线
permeable	['pə:miəbl]	adj.	有浸透性的,能透过的
barrier	['bæriə]	n.	障碍
fence	[fens]	vt.	防护;防卫
organically	[ɔ:'gænikəli]	adv.	器官上地,有机地
microscopic	[maikrə'skɔpik]	adj.	用显微镜可见的,精微的;极小的;极微的
bilayer	[ˌbai'leiə]	n.	双层;双分子层
mosaic	[mə'zeiik]	n.	镶嵌,镶嵌图案,镶嵌工艺;嵌合体
		adj.	嵌花式的,拼成的
plasma	['plæzmə]	n.	血浆
hydrophilic	[haidrə'filik]	adj.	亲水的,喜水的;吸水的
hydrophobic	[ˌhaidrəu'fəubik]	adj.	恐水病的,患恐水病的,不易被水沾湿的
align	[ə'lain]	vt.	排成一行

		vi.	排列
layering	['leiəriŋ]	n.	压条法
cholesterol	[kə'lestərəul]	n.	胆固醇
suspend	[səs'pend]	vt.	吊,悬挂;延缓,暂缓
interior	[in'tiəriə]	adj.	内部的,内的
		n.	内部
gateway	['geitwei]	n.	门,通路,通道
glycolipid	[ˌglaikə'lipid]	n.	糖脂类
attach	[ə'tætʃ]	vt.	加上;连上;附上;依恋;喜欢
term	[tə:m]	vt.	把……称为,称为,叫作
protoplasm	['prəutəplæzəm]	n.	原生质
nucleoplasm	['nju:kliəplæzm]	n.	核原形质,核浆
protistan	[prəu'tistən]	adj.	原生生物的
peptidoglycan	[ˌpeptidəu'glaikæn]	n.	肽聚糖
cellulose	['seljuləus]	n.	植物纤维质;纤维素
lignin	['lignin]	n.	木质素
protist	['prəutist]	n.	原生生物

Phrases

with the exception of...	除……之外
nerve cell	神经细胞
red blood cell	红细胞
the Dead Sea	死海
the Great Salt Lake	大盐湖
the Yellowstone National Park	(美国)黄石国家公园
plasma membrane	质膜
phospholipid bilayer	磷脂双分子层
eukaryotic cell	真核细胞
prokaryotic cell	原核细胞
the majority of	大多数
tail to tail	尾对尾
in exchange for	以……换
integral protein	整合蛋白
gateway protein	通道蛋白

Exercises

【EX. 5】 根据课文内容,回答以下问题

1) What are cells?

2) How many types are all cells divided into? What are they?

3) What is the difference between the prokaryotic cell and the eukaryotic cell?

4) What are eukaryotes?

5) What do eukaryotes and prokaryotes include respectively?

6) Where do methanogens live respectively?

7) Where do extreme halophiles live?

8) Where do extreme thermophiles live?

9) What does the cell membrane function as?

10) Do all living things have cell walls? What about bacteria? What is the most common chemical in the plant primary cell wall?

Reading Material

Text	Notes
The Nucleus and Cytoplasm 　　The nucleus occurs only in eukaryotic cells, and is the location of the majority of different types of nucleic acids. Van Hammerling's experiment showed the role of the nucleus in controlling the shape and features of the cell. Deoxyribonucleic acid, DNA, is the physical carrier of inheritance and with the exception of plastid[1] DNA (cpDNA and mDNA, see below) all DNA	[1]质体

续表

Text	Notes
is restricted to the nucleus. Ribonucleic acid, RNA, is formed in the nucleus by coding off the DNA bases. RNA moves out into the cytoplasm. The nucleolus is an area of the nucleus (usually 2 nucleoli per nucleus) where ribosomes are constructed. The nuclear envelope[2] is a double-membrane structure. Numerous pores[3] occur in the envelope, allowing RNA and other chemicals to pass, but the DNA not to pass. The cytoplasm was defined earlier as the material between the plasma membrane (cell membrane) and the nuclear envelope. Fibrous proteins that occur in the cytoplasm, referred to as the cytoskeleton[4] maintain the shape of the cell as well as anchoring organelles, moving the cell and controlling internal movement of structures. Microtubules function in cell division and serve as a "temporary scaffolding[5]" for other organelles. Actin filaments[6] are thin threads that function in cell division and cell motility. Intermediate filaments are between the size of the microtubules and the actin filaments. Vacuoles[7] are single-membrane organelles that are essentially part of the outside that is located within the cell. The single membrane is known in plant cells as a tonoplast[8]. Many organisms will use vacuoles as storage areas. Vesicles[9] are much smaller than vacuoles and function in transport within and to the outside of the cell. Ribosomes are the sites of protein synthesis. They are not membrane-bound and thus occur in both prokaryotes and eukaryotes. Eukaryotic ribosomes are slightly larger than prokaryotic ones. Structurally the ribosome consists of a small and larger subunit. Biochemically the ribosome consists of ribosomal RNA (rRNA) and some 50 structural proteins. Often ribosomes cluster on the endoplasmic reticulum[10], in which case they resemble a series of factories adjoining a railroad line. Endoplasmic reticulum is a mesh of interconnected membranes that serve a function involving protein synthesis and transport. Rough endoplasmic reticulum (Rough ER) is so-named because of its rough appearance due to the numerous ribosomes that occur along	[2]包膜 [3]孔；毛孔 [4]细胞骨架 [5]脚手架 [6]细肌丝，肌动蛋白丝 [7]液泡 [8]液泡膜 [9]小泡；气泡囊 [10]内质网

Text	Notes
the ER. Rough ER connects to the nuclear envelope through which the messenger RNA (mRNA) that is the blueprint[11] for proteins travels to the ribosomes. Smooth ER lacks the ribosomes characteristic of Rough ER and is thought to be involved in transport and a variety of other functions.	[11]蓝图;计划
Golgi complexes[12] are flattened stacks of membrane-bound sacs. They function as a packaging plant, modifying vesicles from the Rough ER. New membrane material is assembled in various cisternae[13] of the golgi.	[12]高尔基复合体 [13]池;内胞浆网槽
Lysosomes[14] are relatively large vesicles formed by the Golgi. They contain hydrolytic enzymes that could destroy the cell. Lysosome contents function in the extracellular breakdown of materials.	[14]溶酶体
Mitochondria[15] contain their own DNA (termed mDNA) and are thought to represent bacteria-like organisms incorporated into eukaryotic cells over 700 million years ago (perhaps even as far back as 1.5 billion years ago). They function as the sites of energy release (following glycolysis[16] in the cytoplasm) and ATP formation (by chemiosmosis[17]). The mitochondrion has been termed the powerhouse of the cell. Mitochondria are bounded by two membranes. The inner membrane folds into a series of cristae[18], which are the surfaces on which ATP is generated.	[15]线粒体 [16]糖酵解 [17]化学渗透 [18]羽冠,脊
During the 1980s, Lynn Margulis proposed the theory of endosymbiosis[19] to explain the origin of mitochondria and chloroplasts[20] from permanent resident prokaryotes. According to this idea, a larger prokaryote (or perhaps early eukaryote) engulfed[21] or surrounded a smaller prokaryote some 1.5 billion to 700 million years ago.	[19]内共生 [20]叶绿体 [21]卷入,吞没
Instead of digesting the smaller organisms the large one and the smaller one entered into a type of symbiosis known as mutualism[22], wherein both organisms benefit and neither is harmed. The larger organism gained excess ATP provided by the "protomitochondrion[23]" and excess sugar provided by the "protochloroplast[24]", while providing a stable environment and the raw materials the endosymbionts required. This is so strong that now eukaryotic	[22]互惠共生(现象),共栖 [23]原线粒体 [24]原叶绿体

Text	Notes
cells cannot survive without mitochondria (likewise photosynthetic eukaryotes[25] cannot survive without chloroplasts), and the endosymbionts cannot survive outside their hosts[26]. Nearly all eukaryotes have mitochondria. Mitochondrial division is remarkably similar to the prokaryotic methods that will be studied later in this course. Plastids are also membrane-bound organelles that only occur in plants and photosynthetic eukaryotes. Chloroplasts are the sites of photosynthesis[27] in eukaryotes. They contain chlorophyll, the green pigment necessary for photosynthesis to occur, and associated accessory pigments (carotenes[28] and xanthophylls[29]) in photosystems[30] embedded in membranous sacs[31], thylakoids[32] (collectively a stack of thylakoids) are a granum[33] floating in a fluid termed the stroma[34]. Chloroplasts contain many different types of accessory pigments, depending on the taxonomic[35] group of the organism being observed. Like mitochondria, chloroplasts have their own DNA, termed cpDNA. Chloroplasts of green algae[36] (protista) and plants (descendants of some green algae) are thought to have originated by endosymbiosis of a prokaryotic alga similar to living Prochloron (Prochlorobacteria). Chloroplasts of red algae[37] (protista) are very similar biochemically to cyanobacteria[38] (also known as bluegreen bacteria). Endosymbiosis is also invoked for this similarity, perhaps indicating more than one endosymbiotic event occurred.	[25] 光合真核细胞 [26] 寄主,宿主 [27] 光合作用 [28] 胡萝卜素 [29] 叶黄素 [30] 光合体系 [31] 囊;液囊 [32] 类囊体 [33] 叶绿体基粒 [34] 基质 [35] 分类学的;分类的 [36] 绿藻 [37] 红藻 [38] 蓝细菌

Text A 参考译文

有 丝 分 裂

1. 前期:形成有丝分裂装置

染色体浓缩始于 G2 期,在光学显微镜下可见单个浓缩的染色体时,有丝分裂的第一阶段——前期就已经开始了。浓缩的过程在整个前期持续,因此,有些染色体在前期开始时仅仅为微丝,而在这一阶段结束前看起来却相当大。当带有 rRNA 基因的一部分染色体被浓缩时,核糖体 RNA 合成就停止了。

纺锤体的装配。以后将分离姐妹染色单体的微管装置的装配在有丝分裂前期进行。

在动物细胞中,在有丝分裂前期的初期,在 G2 期形成的两个配对的中心粒开始分别向两极移动,在它们之间形成一个叫作纺锤丝的微管轴。同时,细胞中心粒到达细胞的两极,在它们之间建立一座叫做纺锤体的微管桥。在植物细胞中,细胞两极之间也形成一个类似的微管纤维桥,尽管植物细胞内没有中心粒。

在纺锤体的形成过程中,细胞核膜破裂,内质网重新吸收它的成分。在这种情况下,微管纺锤纤维随后完全地跨越细胞,从一极到另一极。纺锤体微管的方向决定了分裂细胞的细胞板的位置;细胞板穿过细胞中央,其方向与纺锤体垂直。

在动物细胞有丝分裂中,当中心粒到达细胞两极时,中心粒向细胞膜伸展放射状排列的微管。这种排列的微管叫做星状体。虽然星状体的功能还没有被完全了解,它可能支撑中心粒离开细胞膜,在纺锤丝消失期间使微管附着点黏稠、坚硬。植物细胞有坚硬的细胞壁,不形成星状体。

将姐妹染色单体拉向两极。每一条染色体拥有两个着丝点,一个着丝点连着一个姐妹染色单体的着丝粒区域。随着前期的继续,第二微管团出现,并从细胞的两极向着丝点方向生长。这些微管将每一对姐妹染色单体着丝点与两极的纺锤体连接起来。因为从两极伸展的微管将着丝点拉向相反方向,它们将一条姐妹染色单体拉向一极,而另一条拉向另一极。这种安排在有丝分裂过程中绝对是关键的,微管配置的任何错误都会导致严重的后果。如着丝点的两边都向同一极移动,将导致姐妹染色单体分离的失败,最终它们在同一子细胞中死亡。

2. 中期:着丝点的排列

有丝分裂的第二阶段——中期是染色体在细胞中心排列的时期。在光学显微镜下观察,染色体沿着细胞内部周围似乎排列成一圈,像地球的赤道带。与通过这个圆圈的纺锤体轴垂直的一个想象的平面被叫做中期板。中期板不是一个真正的结构,而是作为以后细胞分裂轴的标志。根据连接着丝点的着丝粒的微管的位置,所有染色体都排列在中期板上。这是中期结束的标志。染色体的着丝点整齐地排列成圆圈,这个圆圈与细胞两极呈等距离。此时向细胞另一极伸展的微管呈现一种排列并由其形状而得名为纺锤体。

3. 后期和末期:染色单体的分离以及细胞核的重新形成

有丝分裂的所有时期中,后期时间最短,但最具观赏性。这一阶段在着丝点分离时开始。每一个着丝点分裂成两个,两条染色单体彼此分离。所有染色体的着丝点同时分离,但是现在我们还不清楚达到这种同步性的机制。彼此分离的姐妹染色单体迅速地被拉向与它们的着丝粒相连接的细胞两极。在这个过程中,受微管控制的两种形式的移动同时发生。首先,由于完全地固定到相反极上的微管纺锤丝越过彼此向远离细胞中心方向滑动,细胞两极分别移动。因为另一组微管将染色体移向细胞极,染色体也分别移动。如果一个有弹性的细胞膜包围这个细胞,它会显著地被拉长。第二,由于连接着丝点到两极的微管变短,着丝点向两极移动。这种变短的过程不是一种收缩,微管一点都不会变粗。而微管蛋白亚基被微管组织中心从微管的着丝粒端部移开。当更多的亚基被移走时,带有染色单体的微管渐渐地分解,染色单体不断地被牵引到细胞两极。

当姐妹染色单体在细胞分裂后期分离时,复制的染色体组(有丝分裂的主要元素)的精确分割就完成了。在细胞分裂末期,由于微管破裂成能被用于构建子细胞骨架的微管蛋白单体,纺锤体装置解体了。每一套姐妹染色单体周围形成一个核膜,现在染色单体可以被称作染色体,因为每一条都有自己的着丝点。染色体很快开始重新解旋,成为允许基因表达的更长的形式。导致核仁重新出现的 rRNA 基因是早期表达基因之一。

Text B 参考译文

细　胞

A. 细胞和有机体的分类

细胞是生物机体的基本单元,但一些结构与功能与细胞不同的病毒除外。虽然神经细胞与红细胞的外观完全不同,但它们的结构在本质上是一样的。甚至植物细胞和动物细胞的整个结构都极其相似。

所有的细胞可分为两类:原核细胞和真核细胞。原核细胞没有细胞核,而真核细胞有一个细胞核。

真核生物是由真核细胞组成的有机体,包括原生生物、真菌、动物和植物。原核生物包括原始细菌和真细菌,它们都是单细胞生物。

最近,"原始细菌"已被列到"细菌"范畴之外,因为它们与普通的细菌完全不同。根据新的分类,原核生物分为古细菌(*archaea*)和细菌(*bacteria*),这里的"古细菌(*archaea*)"等同于"原始细菌(*archaebacteria*)","细菌(*bacteria*)"等同于"真细菌(*eubacteria*)"。

古细菌生活在极端的环境中,它们也许可被分为三个种群:

甲烷细菌在厌氧的环境(如沼泽)中生存,它们产生甲烷,而且不能暴露在氧气中。

极端嗜盐生物生活在盐度很高的环境中,如死海和大盐湖。

极端嗜热生物生活在高温、富硫和低 pH 值的环境中,如温泉、锅炉和黄石国家公园(美国)中的火山喷气孔。

B. 基本细胞组分

所有的细胞都包含有细胞质、质膜和 DNA。

细胞质是细胞的黏性内容物,包含有蛋白质、核糖体、代谢物和离子,其中核糖体是蛋白质合成的场所。

质膜是包围细胞质的细胞膜,由磷脂双分子层以及蛋白质和碳水化合物组成。磷脂双分子层也是其他生物膜的基本成分。

DNA(脱氧核糖核酸)是遗传物质。一个真核细胞具有几个 DNA 分子,这些分子位于有膜的细胞器——细胞核和线粒体内。一个原核生物只有一个 DNA 分子,它与细胞质之间没有特殊的界线。

C. 细胞膜和细胞壁

　　细胞膜的功能是作为一种半渗透的屏障,允许极少数的分子穿过,它把大部分的细胞器产生的化学物质隔离在细胞内部。用电子显微镜对细胞膜的观察促进了对脂质双分子层结构模型(也称为液态镶嵌模型)的研究。这种模型中最常见的分子是磷脂,它有一个极性(亲水)头和两个非极性(疏水)尾。这些磷脂呈尾对尾状排列,这样非极性区域在膜内外表面的亲水头之间形成了一个疏水区。自从有一种被称为冷冻断裂的电子显微技术可以分开这一分层以后,这种分层就被定义为双分子层。

　　胆固醇是细胞膜的另一种重要成分,它深入到内部(尾对尾)的疏水区。大多数细菌的细胞膜不具有胆固醇。

　　蛋白质悬在内层中,但它们更多的亲水区"伸到"细胞内部和外部。这些蛋白质有通道的功能和交换作用,允许特定的分子进出细胞。这种整合蛋白有时候被称为通道蛋白。膜的外表面通常富含糖脂,糖脂的疏水尾端深入到膜的疏水区内,它们的头端暴露在细胞外。这些糖脂与附着在整合蛋白上的碳水化合物一起,被认为有自我识别的功能。

　　细胞的内容物(包括化学物质和细胞器)被称为原生质体,而且可被进一步细分为细胞质(所有除核所包含的物质外的原生质体)和核质(细胞核中的所有的原料、胞浆和DNA等)。

　　不是所有的生物都有细胞壁,特别是动物和许多与动物较相似的原生生物。细菌具有细胞壁,这些细胞壁含有肽聚糖。植物细胞的细胞壁由多种化学物质组成,其中纤维素是初生细胞壁中最常见的化学物质。有些植物细胞的次生壁中也含有木质素及其他的化学物质。细胞壁位于质膜的外面。通过胞间连丝的连接,细胞能穿透相互之间厚实的细胞壁进行化学交换。真菌类和许多原生生物都具有细胞壁,但它们不含纤维素,而具有多种化学物质(如真菌中的壳聚糖)。

Unit 3

Text A

The Modern View of the Genes

While Mendel discussed traits, we now know that genes are segments of the DNA that code for specific proteins. These proteins are responsible for the expression of the phenotype. The basic principles of segregation and independent assortment as worked out by Mendel are applicable even for sex-linked traits.

Codominant alleles

Codominant alleles occur when rather than expressing an intermediate phenotype, the heterozygotes express both homozygous phenotypes. An example is in human ABO blood types, the heterozygote AB type manufactures antibodies to both A and B types. Blood Type A people manufacture only anti-B antibodies, while type B people make only anti-A antibodies. Codominant alleles are both expressed. Heterozygotes for codominant alleles fully express both alleles. Blood type AB individuals produce both A and B antigens. Since neither A nor B is dominant over the other and they are both dominant over O they are said to be codominant.

Incomplete dominance

Incomplete dominance is a condition when neither allele is dominant over the other. The condition is recognized by the heterozygotes expressing an intermediate phenotype relative to the parental phenotypes. If a red flowered plant is crossed with a white flowered one, the progeny will all be pink. When pink is crossed with pink, the progeny are 1 red, 2 pink, and 1 white.

Flower color in snapdragons is an example of this pattern. Cross a true-breeding red strain with a true-breeding white strain and the F1 are all pink (heterozygotes). Self-fertilize the F1 and you get an F2 ratio of 1 red : 2 pink : 1 white. This would not happen if true blending had occurred (blending cannot explain traits such as red or white skipping a generation and pink flowers crossed with pink flowers should produce only pink flowers).

Multiple alleles

Many genes have more than two alleles (even though any one diploid individual can only have at most two alleles for any gene), such as the ABO blood groups in humans, which are an example of multiple alleles. Multiple alleles result from different mutations of the same gene. Coat color in rabbits is determined by four alleles. Human ABO blood types are determined by alleles A, B, and O. A and B are codominants which are both dominant over O. The only possible genotype for a type O person is OO. Type A people have either AA or AO genotypes. Type B people have either BB or BO genotypes. Type AB people have only the AB (heterozygous) genotype. The A and B alleles of gene I produce slightly different glycoproteins (antigens) that are on the surface of each cell. Homozygous A individuals have only the A antigen, homozygous B individuals have only the B antigen, homozygous O individuals produce neither antigen, while a fourth phenotype (AB) produces both A and B antigens.

Interactions among genes

While one gene may make only one protein, the effects of those proteins usually interact, for example widow's peak may be masked by expression of the baldness gene. Novel phenotypes often result from the interactions of two genes, as in the case of the comb in chickens. The single comb is Epistasis.

Epistasis is the term appl produced only by the rrpp genotype. Rose comb (b) results from R_pp. (_ can be either R or r). Pea comb (c) results from rrP_. Walnut comb, a novel phenotype, is produced when the genotype has at least one dominant of each gene (R_P_). Bateson reported a different phenotypic ratio in sweet pea than could be explained by simple Mendelian inheritance. This ratio is 9 : 7 instead of the 9 : 3 : 3 : 1. one would expect of a dihybrid cross between heterozygotes. Of the two genes (C and P), when either is homozygous recessive (cc or pp) that gene is epistatic to (or hides) the other. To get purple flowers one must have both C and P alleles present.

Environment and Gene Expression

Phenotypes are always affected by their environment. In buttercup (*Ranunculus peltatus*), leaves below water-level are finely divided and those above water-level are broad, floating, photosynthetic leaf-like leaves. Siamese cats are darker on their extremities, due to temperature effects on phenotypic expression. Expression of phenotype is a result of interaction between genes and environment. Siamese cats and Himalayan rabbits both have dark colored fur on their extremities. This is caused by an allele that controls pigment production being able only to function at the lower temperatures of those extremities. Environment determines the phenotypic pattern of expression.

New Words

gene	[dʒi:n]	n.	遗传因子,基因
trait	[treit]	n.	性状,特性
segment	['segmənt]	n.	体节,环节;片段
code	[kəud]	n.	遗传(密)码
expression	[iks'preʃən]	n.	表达
phenotype	['fi:nətaip]	n.	显型,表现型
principle	['prinsəpl]	n.	法则,原则,原理
segregation	[ˌsegri'geiʃən]	n.	分离,隔离
independent	[indi'pendənt]	adj.	独立自主的,不受约束的
assortment	[ə'sɔ:tmənt]	n.	分类,分配,搭配
heterozygote	[hetərəu'zaigəut]	n.	异质接合体,异形接合体,杂合体,杂合子
homozygous	[ˌhɔmə'zaigəs]	adj.	同型结合的,纯合子的
manufacture	[ˌmænju'fæktʃə]	vt.	制造,加工
		n.	制造,制造业,产品
antibody	['æntiˌbɔdi]	n.	抗体
codominant	[kəu'dɔminənt]	adj.	等显性的,共显性的
allele	[ə'li:l]	n.	等位基因
antigen	['æntidʒən]	n.	抗原
dominant	['dɔminənt]	adj.	显性的,优势的
dominance	['dɔminəns]	n.	优势;显性
incomplete	[ˌinkəm'pli:t]	adj.	不完全的,不完善的
progeny	['prɔdʒini]	n.	后裔
pink	[piŋk]	adj.	粉红色的
		n.	粉红色
snapdragon	['snæpdrægən]	n.	金鱼草
strain	[strein]	n.	系,品系,株
diploid	['diplɔid]	adj.	双倍的
mutation	[mju:'teiʃən]	n.	突变
rabbit	['ræbit]	n.	兔
genotype	['dʒenətaip]	n.	基因型
glycoprotein	[ˌglaikəu'prəuti:n]	n.	糖蛋白

homozygous	[ˌhɔməˈzaigəus]	n.	纯合子的,同型结合的
novel	[ˈnɔvəl]	adj.	新的
comb	[kəum]	n.	鸡冠
epistasis	[iˈpistəsis]	n.	上位显性
dihybrid	[daiˈhaibrid]	n.	双杂交
recessive	[riˈsesiv]	adj.	隐性的
		n.	隐性性状
purple	[ˈpəːpl]	adj.	紫色的
		n.	紫色
buttercup	[ˈbʌtəkʌp]	n.	[植]毛茛科
photosynthetic	[ˌfəutəusinˈθetik]	adj.	光合作用的,光合的
extremity	[iksˈtremiti]	n.	末端,极端
pigment	[ˈpigmənt]	n.	色素,颜料
phenotypic	[ˈfiːnəˌtipik]	adj.	表型的
epistatic	[ˌepəˈstætik]	adj.	上位的,强性的

Phrases

be responsible for	对……负责
work out	算出
be applicable for	可应用到……;对……很合适
codominant alleles	显性等位基因
intermediate phenotype	中间表现型,中间显型
homozygous phenotype	纯合表现型,纯合显型
blood type	血型
parental phenotype	亲本显型
multiple allele	复等位基因
at most	最多;最大
result from	由……而造成
pea comb	豆冠
walnut comb	胡桃冠
widow's peak	额头的 V 型发尖
coat color	毛色
be determined by	由……来决定
be masked by	被……隐蔽
at least	至少
interfere with	干扰
Siamese cat	暹罗猫

Notes

[1] Self-fertilize the F1 and you get an F2 ratio of 1 red : 2 pink : 1 white.

本句所使用的句型是：祈使句＋and＋句子，等于 If 从句＋主句。本句可改写为：If you self-fertilize the F1, you will get an F2 ratio of 1 red : 2 pink : 1 white. 请看下例：

Leave a basin of water out of the room in cold winter and it will become a block of ice.

在寒冷的冬天把一盆水放在户外，它就会结成一块冰。

Use your brain and you will find a way.

动动脑筋，你就会想出办法来。

[2] Many genes have more than two alleles (even though any one diploid individual can only have at most two alleles for any gene), such as the ABO blood groups in humans, which are an example of multiple alleles.

本句中，even though any one diploid individual can only have at most two alleles for any gene 是一个让步状语从句，修饰主句的谓语 have，which are an example of multiple alleles 是一个非限定性定语从句，对 the ABO blood groups in humans 做补充说明。

[3] While one gene may make only one protein, the effects of those proteins usually interact, for example widow's peak may be masked by expression of the baldness gene.

本句中，While one gene may make only one protein 是一个让步状语从句，修饰主句的谓语 interact。请看下例：

While the grandparents love the children, they are strict with them.

虽然祖父母们都爱他们的孩子，但却对他们要求严格。

[4] This is caused by an allele that controls pigment production being able only to function at the lower temperatures of those extremities.

本句中，that controls pigment production being able only to function at the lower temperatures of those extremities 是一个定语从句，修饰和限定 an allele。

Exercises

【EX. 1】 根据课文内容，回答以下问题

1) What are genes? And what are these proteins responsible for?

2) When do codominant alleles occur?

3) What do blood type A and type B people manufacture?

4) What is incomplete dominance?

5) What do multiple alleles result from?

6) What is the possible genotype for a type O person? What about type A, type B and type AB people?

7) What do novel phenotypes often result from?

8) What is Epistasis?

9) What are phenotypes always affected by?

10) What is expression of phenotype?

【EX. 2】 根据下面的英文解释,写出相应的英文词汇

英 文 解 释	词 汇
The observable physical or biochemical characteristics of an organism, as determined by both genetic makeup and environmental influences.	
An organism that has different alleles at a particular gene locus on homologous chromosomes.	
Having the same alleles at a particular gene locus on homologous chromosomes.	
A sudden structural change within a gene or chromosome of an organism resulting in the creation of a new character or trait not found in the parental type.	
Of, relating to, or designating an allele that does not produce a characteristic effect when present with a dominant allele.	
A genetically determined characteristic or condition.	
A hereditary unit that occupies a specific location on a chromosome and determines a particular characteristic in an organism.	

续表

英　文　解　释	词　汇
A substance that when introduced into the body stimulates the production of an antibody.	
Any of a group of conjugated proteins that contain a carbohydrate as the nonprotein component.	
The hybrid of parents that differ at only two gene loci, for which each parent is homozygous with different alleles.	

【EX. 3】 把下列句子翻译为中文

1) Parents contribute a unique subset of alleles to each of their nonidentical twin offspring.

2) The phenotype of a given individual for a particular gene depends on which alleles it carries (its genotype) and whether the alleles are dominant, recessive, or codominant.

3) The role of genes in determining the phenotype for a quantitative trait is estimated by calculating the heritability of the trait.

4) Although Gregor Mendel's work was meticulous and well documented, his discoveries, reported in the 1860s, were ignored until decades later.

5) Dominance is sometimes not complete, since both alleles in a heterozygous organism may be expressed in the phenotype.

6) Genes are expressed in the phenotype as polypeptides (proteins).

7) Certain hereditary diseases in humans had been found to be caused by the absence of certain enzymes. These observations supported the one-gene, one-polypeptide hypothesis.

8) Mutations in DNA are often expressed as abnormal proteins. However, the result may not be easily observable phenotypic changes. Some mutations are detectable only under certain conditions.

9) Chromosomal mutations (deletions, duplications, inversions, or translocations) involve large regions of a chromosome.

10) Mutations can be spontaneous or induced. Spontaneous mutations occur because of instabilities in DNA or chromosomes.

【EX. 4】 把下列短文翻译为中文

DNA was proven as the hereditary material and Watson et al. had deciphered its structure. What remained was to determine how DNA copied its information and how that was expressed in the phenotype. Matthew Meselson and Franklin W. Stahl designed an experiment to determine the method of DNA replication. Three models of replication were considered likely.

1) Conservative replication would somehow produce an entirely new DNA strand during replication.

2) Semiconservative replication would produce two DNA molecules, each of which was composed of one-half of the parental DNA along with an entirely new complementary strand. In other words the new DNA would consist of one new and one old strand of DNA. The existing strands would serve as complementary templates for the new strand.

3) Dispersive replication involved the breaking of the parental strands during replication, and somehow, a reassembly of molecules that were a mix of old and new fragments on each strand.

Text B

The Human Karyotype

There are 44 autosomes and 2 sex chromosomes in the human genome, for a total of 46. Karyotypes are pictures of homologous chromosomes lined up together during Metaphase I of meiosis. The chromcsome micrographs are then arranged by size and pasted onto a sheet.

1. Human chromosomal abnormalities

A common abnormality is caused by nondisjunction, the failure of replicated chromosomes to segregate during Anaphase II. A gamete lacking a chromosome cannot produce a viable embryo. Occasionally a gamete with $n+1$ chromosomes can produce a viable embryo.

In humans, nondisjunction is most often associated with the 21st chromosome, producing a disease known as Down's syndrome (also referred to as trisomy 21). Sufferers of Down's syndrome suffer mild to severe mental retardation, short stocky body type, large tongue leading to speech difficulties, and (in those who survive into middle-age), a propensity to develop Alzheimer's Disease. Ninety-five percent of Down's cases result from nondisjunction of chromosome 21. Occasional cases result from a translocation in the chromosomes of one parent. Remember that a translocation occurs when one chromosome (or a fragment) is transferred to a non-homologous chromosome. The incidence of Down's Syndrome increases with age of the mother, although 25% of the cases result from an extra chromosome from the father.

Sex-chromosome abnormalities may also be caused by nondisjunction of one or more sex chromosomes. Any combination (up to XXXXY) produces maleness. Males with more than one X are usually underdeveloped and sterile. XXX and XO women are known, although in most cases they are sterile. What meiotic difficulties might a person with Down's syndrome or extra sex-chromosomes face?

Chromosome deletions may also be associated with other syndromes such as Wilm's tumor.

Prenatal detection of chromosomal abnormalities is accomplished chiefly by amniocentesis. A thin needle is inserted into the amniotic fluid surrounding the fetus (a term applied to an unborn baby after the first trimester). Cells withdrawn have been sloughed off by the fetus, yet they are still fetal cells and can be used to determine the state of the fetal chromosomes, such as Down's Syndrome and the sex of the baby after a karyotype has been made.

2. Human Allelic Disorders (Recessive)

The first Mendelian trait in humans was described in 1905 (brachydactly) by Dr. Farabee. Now more than 3500 human genetic traits are known.

Albinism, the lack of pigmentation in skin, hair, and eyes, is also a Mendelian human trait. Homozygous recessive (aa) individuals make no pigments, and so have face, hair, and eyes that are white to yellow. For heterozygous parents with normal pigmentation (Aa), two different types of gametes may be produced: A or a. From such a cross 1/4 of the children could be albinos. The brown pigment melanin cannot be made by albinos. Several mutations may cause albinism: 1) the lack of one or another enzyme along the melanin-producing pathway; or 2) the inability of the enzyme to enter the pigment cells and convert the amino acid tyrosine into melanin.

Phenylketonuria (PKU) is recessively inherited disorder whose sufferers lack the ability to synthesize an enzyme to convert the amino acid phenylalanine into tyrosine. Individuals homozygous recessive for this allele have a buildup of phenylalanine and abnormal breakdown products in the urine and blood. The breakdown products can be harmful to developing nervous systems and lead to mental retardation. 1 in 15,000 infants suffers from this problem. PKU homozygotes are now routinely tested for in most states. If you look closely at a product containing Nutra-sweet artificial sweetener, you will see a warning to PKU sufferers since phenylalanine is one of the amino acids in the sweetener. PKU sufferers are placed on a diet low in phenylalanine, enough for metabolic needs but not enough to cause the buildup of harmful intermediates.

Tay-Sachs Disease is an autosomal recessive resulting in degeneration of the nervous system. Symptoms manifest after birth. Children homozygous recessive for this allele rarely survive past five years of age. Sufferers lack the ability to make the enzyme N-acetyl-hexosaminidase, which breaks down the GM2 ganglioside lipid. This lipid accumulates in lysosomes in brain cells, eventually killing the brain cells. Although rare in the general population (1 in 300,000 births), it was (until recently) higher (1 in 3600 births) among Jews of eastern central European descent. One in 28 American Jews is thought to be a carrier, since 90% of the American Jewish population emigrated from those areas in Europe. Most Tay-Sachs babies born in the US are born to non-Jewish parents, who did not undergo testing programs that most US Jewish prospective parents had.

Sickle-cell anemia is an autosomal recessive. Nine-percent of US blacks are heterozygous, while 0.2% are homozygous recessive. The recessive allele causes a single amino acid substitution in the beta chains of hemoglobin. When oxygen concentration is low, sickling of cells occurs. Heterozygotes make enough "good beta-chain hemoglobin" that they do not suffer as long as oxygen concentrations remain high, such as at sea-level.

3. Human Allelic Disorders (Dominant)

Autosomal dominants are rare, although they are (by definition) more commonly expressed.

Achondroplastic dwarfism occurs, even though sufferers have reduced fertility.

Huntington's disease (also referred to as Woody Guthrie's disease, after the folk singer who died in the 1960s) is an autosomal dominant resulting in progressive destruction of brain cells. If a parent has the disease, 50% of the children will have it (unless that parent was homozygous dominant, in which case all children would have the disease). The disease usually does not manifest until after age 30, although some instances of early onset phenomenon are reported among individuals in their twenties.

Polydactla is the presence of a sixth digit. In modern times the extra finger has been cut off at birth and individuals do not know they carry this trait. One of the wives of Henry VIII had an extra finger. In certain southern families the trait is also more common. The extra digit is rarely functional and definitely causes problems buying gloves, let alone fitting them on during a murder trial.

New Words

karyotype	['kæriətaip]	n.	染色体组型
autosome	['ɔːtəsəum]	n.	常染色体
genome	['dʒiːnəum]	n.	基因组,染色体组
meiosis	[maiˈəusis]	n.	减数分裂
nondisjunction	['nɔndisˈdʒʌŋkʃən]	n.	不分离
gamete	['gæmiːt]	n.	配子,接合体
propensity	[prəˈpensiti]	n.	倾向
translocation	[ˌtrænsləuˈkeiʃən]	n.	易位
sterile	['sterail]	adj.	不育的,不结果的
meiotic	[maiˈɔtik]	adj.	减数分裂的
prenatal	[priːˈneitəl]	adj.	产前的,出生前的
tumor	['tjuːmə]	n.	肿瘤,肿块
amniocentesis	[ˌæmniəusenˈtiːsis]	n.	羊水诊断
fetus	['fiːtəs]	n.	胎,胎儿
trimester	[traiˈmestə]	n.	三个月
albinism	['ælbənizəm]	n.	白化病,皮肤变白症
brachydactyly	[ˌbrækiˈdæktili]	n.	短指,短趾
pigmentation	[ˌpigmənˈteiʃən]	n.	染色,着色,色素沉淀
melanin	['melənin]	n.	黑色素
albinos	[ælˈbiːnəu]	n.	白化病者;白化变种
retardation	[ˌriːtɑːˈdeiʃən]	n.	阻滞,迟缓;迟钝
phenylalanine	[ˌfenəlˈæliːniːn]	n.	苯丙氨酸
intermediate	[ˌintəˈmiːdjət]	n.	中间体,媒介物,媒介
		adj.	中间的
accumulate	[əˈkjuːmjuleit]	v.	积聚,堆积
lysosome	['laisəsəum]	n.	溶酶体
autosomal	[ˌɔːtəuˈsəuməl]	adj.	常染色体的
hemoglobin	[ˌhiːməuˈgləubin]	n.	血红蛋白,血色素
heterozygote	[hetərəuˈzaigəut]	n.	异质接合体,异形接合体
achondroplastic	[əˈkɔndrəˌplæsti]	adj.	软骨发育不全的
dwarfism	['dwɔːfizəm]	n.	矮小症,侏儒症
polydactly	[ˌpɔliˈdæktili]	n.	多指(或多趾)畸形症

Phrases

sex chromosome	性染色体
homologous chromosome	同源染色体
amniotic fluid	羊膜水
slough off	脱落
amino acid tyrosine	酪氨酸
phenylketonuria (PKU)	苯丙酮酸尿
amino acid phenylalanine	苯丙氨酸
ganglioside lipid	神经节苷脂
sickle-cell anemia	镰刀形细胞贫血
allelic disorder	等位基因异常
let alone	不管

Exercises

【EX. 5】 根据课文内容,回答以下问题

1) How many chromosomes are there in human genome? What are they?

2) What is a common abnormality caused by?

3) What may cause sex-chromosome abnormalities?

4) How many human genetic traits are known?

5) What is albinism?

6) What is phenylketonuria?

7) What is Tay-Sachs Disease?

8) What is sickle-cell anemia?

9) What is Huntington's disease?

10) What is polydactly?

Reading Material

Text	Notes
Polygenic Inheritance and Chromosomes Polygenic inheritance[1] is a pattern responsible for many features that seem simple on the surface. Many traits such as height, shape, weight, color, and metabolic rate are governed by the cumulative[2] effects of many genes. Polygenic traits are not expressed as absolute or discrete characters, as was the case with Mendel's pea plant traits. Instead, polygenic traits are recognizable by their expression as a gradation of small differences (a continuous variation). The results form a bell shaped curve, with a mean value and extremes in either direction. Height in humans is a polygenic trait, as is color in wheat kernels. Height in humans is not discontinuous[3]. If you line up the entire class a continuum of variation is evident, with an average height and extremes in variation (very short [vertically challenged] and very tall [vertically enhanced]). Traits showing continuous variation are usually controlled by the additive effects of two or more separate gene pairs. Linkage[4] occurs when genes are on the same chromosome. Remember that sex-linked genes are on the X chromosome, one of the sex chromosomes. Linkage groups are invariably the same number as the pairs of homologous chromosomes an organism possesses. Recombination[5] occurs when crossing-over has broken linkage groups, as in the case of the genes for wing size and body color that Morgan studied. Chromosome mapping was originally based on the frequencies of recombination between alleles. Since mutations can be induced (by radiation or chemicals), Morgan and his coworkers were able to cause new alleles to form by subjecting fruit flies to mutagens[6] (agents of mutation, or mutation generators). Genes are located on specific regions of a certain chromosome, termed the gene locus[7] (plural: loci). A gene therefore is a specific segment of the DNA molecule. Alfred Sturtevant, while an undergraduate student in Morgan's lab, postulated[8] that crossing-over[9] would be less common	[1]多基因遗传 [2]累积的；渐增的 [3]不连续的，间断的 [4]连锁 [5]重组 [6]诱变因素，诱变剂 [7]基因座 [8]假定 [9]交换

Text	Notes
between genes adjacent to each other on the same chromosome and that it should be possible to plot the sequence of genes along a fruit fly chromosome by using crossing-over frequencies. Distances on gene maps are expressed in map units (one map unit = 1 recombinant per 100 fertilized eggs; or a 1% chance of recombination). The map for *Drosophila melanogaster* chromosomes is well known. Note that eye color and aristae length are far apart, as indicated by the occurrence of more recombinants (crossing-overs) between them, while wing length is closer to eye shape (as indicated by the low frequency of recombination between these two features). Chromosome Abnormalities Chromosome abnormalities include inversion[10], insertion[11], duplication, and deletion[12]. These are types of mutations. Since DNA is information, and information typically has a beginning point, an inversion would produce an inactive or altered protein. Likewise deletion or duplication will alter the gene product. Deletion A B C D E F G → A B E F G + C D (lost) Duplication and deletion of homologous chromosomes A B C D E F G → A B E F G A B C D E F G → A B C D C D E F G Inversion A B C D E F G → A B E D C F G Reciprocal translocation between nonhomologous chromosomes A B C D E F G → A B L M N O H I J K L M N O → H I J K C D E F G Chromosomal mutations.	[10]倒位 [11]插入 [12]缺失

Text A 参考译文

基因的现代观点

虽然孟德尔已讨论了性状，但现在我们知道基因是编码特定蛋白质的 DNA 片断。这些蛋白质负责表现型的表达。分离规律和自由组合规律由孟德尔提出，也可应用于伴性性状遗传。

显性等位基因

显性等位基因出现在当杂合子表达两个纯合表现型时，而不是在表达中间表现型时。例如人类的 ABO 血型，杂合子 AB 型既产生 A 型抗体，也产生 B 型的抗体；A 血型的人只产生抗 B 抗体，而 B 血型的人只产生抗 A 抗体。共显性等位基因都被表达，共显性等位基因的杂合子完全表达了两个等位基因。因为 A 对于 B 或者 B 对于 A 来说都不是显性的；A 或 B 对于 O 型来说，它们二者都是显性的，所以 AB 血型的个体既产生 A 抗原，同时又产生 B 抗原。

不完全显性

不完全显性是当两个等位基因中任何一个基因相对于另一个基因都不是显性时的一种情况。杂合子相对于亲本显型表现出一种中间显型时，可确认为这种情况。如果一株开红花的植株与一株开白花的植株杂交，那么后裔植株开的花都将是粉红色的。当粉红色后裔自交时，下一代开的花的颜色比率将是 1(红)：2(粉红)：1(白)。

金鱼草花的颜色就是这种情况的一个例子。让真实遗传的(纯合)红色品系和真实遗传的(纯合)白色品系杂交，F1 代都是粉红色(杂合子)。F1 代之间进行自体授粉，得到的 F2 代比率为：1(红)：2(粉红)：1(白)。假如是真正的混合色种，那么这种情况可能不会发生(真正混合色种的下一代，不会表达为红色或白色性状；粉红色花之间杂交应当只会产生粉红色花)。

复等位基因

许多基因都有两个以上的等位基因(尽管任何一个二倍体个体的所有基因最多只有两个等位基因)，比如人类的 ABO 血型，是复等位基因的一个例子。复等位基因由同一个基因的不同突变产生。兔子的毛色是由四个等位基因决定的。人类的 ABO 血型是由 A、B 和 O 等位基因决定的。A 和 B 是共显性，它们相对于 O 是显性的。一个血型是 O 型的人，他的基因型惟一可能的就是 OO；A 血型的人的基因型既可能是 AA，也可能是 AO；B 血型的人的基因型既可能是 BB，也可能是 BO；AB 型则只有 AB(杂合的)基因型。基因 I 中的 A 和 B 等位基因产生的细胞表面上的糖蛋白(抗原)有所不同。纯合子 A 个体只有 A 抗原，纯合子 B 个体只有 B 抗原，纯合子 O 个体则不产生 A 或 B 抗体中任意一种，同时第四种表现型(AB)既产生 A 抗原也产生 B 抗原。

基因互作

尽管一个基因可能只产生一种蛋白,但这些蛋白的作用通常是相互影响的,如人额头的V型发尖可能掩蔽于脱发基因的表达中。新的表现型经常是由两个基因相互作用而产生,鸡的鸡冠就是这种情况,单冠是上位显性。

上位显性的术语是 appl,它只由 rrpp 基因型产生。玫瑰冠由 R_pp 产生(_既可以是 R 也可以是 r)。豆冠(c)由 rrP_产生。胡桃冠是一种新的表型,当基因型中每一个基因至少有一个是显性(R_P_)时就会产生这种表型。Bateson 报道了甜豌豆表型的一种不同的比率,这个现象超出了简单的孟德尔遗传法则所能解释的范围。这个比率是9∶7,而不是可用来预测杂合子之间杂交产生的双因子杂合子比率的9∶3∶3∶1。对于这两个基因(C 和 P),当任一对基因是纯合隐性(cc 或 pp)时,这个基因相对于另一个基因是上位的(或隐藏了另一个基因)。要想得到紫色的花,必须既要有 C 等位基因又要有 P 等位基因。

环境和基因表达

表现型总是受到环境的影响,毛茛科植物(*Ranunculus peltatus*)中,处于水面以下的叶子是细条状的,而水面以上的叶子是较宽、漂浮的,为有光合作用的叶状的叶子。暹罗猫的足端的毛色较暗,这是温度对表现型的表达影响的结果。表现型的表达是基因与环境相互作用的结果。暹罗猫和喜马拉雅野兔的足端毛色都较暗。这是由于控制色素产生的等位基因只在温度较低的足端才表达所造成的。环境决定了要表达的表现型式样。

Text B 参考译文

人类的染色体组型

人类基因组共有46条染色体,其中包括44条常染色体和2条性染色体。染色体组型(核型)是减数分裂中期Ⅰ期时排列在一起时的全部同源染色体的图像。染色体的显微图像是按染色体的大小排列并被粘贴到一张纸上。

1. 人类染色体畸变

一种常见的畸变是由不分离引起的,不分离是指细胞分裂后期Ⅱ期染色体复制后不能分离。缺少一条染色体的配子不能发育成活的胚胎。有时候一个具有 $n+1$ 染色体的配子可发育成一个活的胚胎。

在人类当中,最常见的不分离现象是与第21对染色体有关的一种被称为唐氏综合征(唐氏先天愚症)的疾病(也被称作三体性21)。唐氏先天愚症患者有不同程度的智力障碍、体型粗矮,及因大舌头而引起的说话困难等症状,而且(在中年人中),有发展成 Alzheimer 病(阿尔茨海默病)的趋势。95%的唐氏综合征是由于第21对染色体的不分离所引起的,偶有一些病例是由双亲中一方的染色体移位引起。记住移位是发生在当一条染色体(或染色体片断)被转移到一条非同源染色体时。尽管有25%的唐氏综合征病例是由来自父亲染色体增加而引起的,但发生唐氏综合征的发病率随着母亲年龄的增长而提高。

性染色体的畸变也可能是由一个或多个性染色体不分离所引起的。任何的组合（直到XXXXY）都产生雄性。有1个以上X染色体的雄性通常发育不完全而且是不育的。已知有些女性的基因型是XXX和XO,尽管她们在大多数情况下是不育的。一个唐氏综合征或有额外染色体患者面对的到底是什么样的减数分裂困难？

染色体缺失与其他的综合病症相关,比如威耳姆肿瘤。

产前染色体异常检测主要是通过羊水诊断来进行。一根细针伸入到胎儿（这里的胎儿指的是孕三个月之后到出生前的胎儿）周围羊膜水中。利用被胎儿丢弃的脱落细胞制作的染色体组型可确定胎儿染色体的情况,比如,可确定胎儿有无唐氏综合征以及胎儿的性别情况。

2. 人类等位基因异常（隐性）

1905年,Farabee博士描述了人类的第一个孟德尔性状（短指）。现在已知有3500个以上的人类遗传性状。

白化病,表现为皮肤、头发和眼睛缺少色素,也是人类的一种孟德尔性状。隐性纯合子(aa)个体不能产生色素,因而脸、头发和眼睛呈现出白色或黄色。对于具有正常色素的杂合子(Aa)亲本,可能会产生两种不同类型的配子：A或a。杂交后就有1/4的小孩是白化病患者。称为黑色素的褐色色素在白化病患者身上不能产生。有几种突变可能导致白化病：①黑色素生成途径中缺乏一种酶或另一种酶；②进入到色素细胞中并将酪氨酸转化成黑色素的酶失活。

苯丙酮酸尿症（PKU）是一种隐性遗传异常疾病,患者缺乏合成一种酶的能力,这种酶可以将苯丙氨酸转化为酪氨酸。这种纯合隐性基因型的个体在尿和血液中有苯丙氨酸及其他异常分解产物的沉积。这种分解产物对神经系统的发育有害,而且会导致智力缺陷。每15 000个婴儿中就有1个会有这样的问题。现在,在多数地区,PKU纯合体为常规检测项目。如果细读某种含有Nutra-sweet人工甜味剂的产品说明书,你会看到一项针对PKU患者的警告,因为这种人工甜味剂中有一种氨基酸是苯丙氨酸。PKU患者的膳食要求苯丙氨酸含量较低,在满足代谢的需要同时,不会在体内累积有害的中间产物。

泰-萨二氏病（家族性黑蒙痴呆症）是一种常染色体隐性基因引起的神经系统恶化症。在出生后症状就会出现。有这种纯合隐性基因型的儿童很少能活过5岁。患者缺乏合成N-乙酰氨基己糖苷酶的能力,这种酶能分解神经节苷脂GM2。这种脂肪(GM2)在脑细胞中的溶酶体内积累,最终会引起脑细胞死亡。虽然在总人口中发病率很低（300 000婴儿中会出现1例）,但在中东欧系的犹太人中（一直到现在）发病的几率却较高（3600婴儿中会出现1例）。由于90%的美国犹太人是欧洲地区的移民,每28个美国犹太人就有一个人被认为是这种基因的携带者。大部分出生在美国的家族性黑蒙痴呆症病患婴儿的父母是非犹太人,这些非犹太人没有像大部分的美国犹太父母一样经过检查。

镰刀型细胞贫血是一种常染色体的隐性性状。9%的美国黑人具有杂合子,同时有0.2%的美国黑人具有隐性纯合子。隐性基因导致血色素β链中一个氨基酸被替换。当氧浓度较低时,就会产生镰刀形血细胞。杂合体产生足够多的"具有良好β链的血色素",这样只要氧的浓度维持在较高水平,如在海平面上,就不会得病。

3. 人类等位基因的异常(显性)

常染色体显性性状是很少见的,虽然他们的性状更易于被表达出来(在定义上)。

软骨营养障碍性矮小症患者的繁殖率较低,但还是会发生这种病。

亨廷顿病(20世纪60年代,一位民间歌手Woody Guthrie死于这种病后,也被称为Woody Guthrie病)是一种常染色体引起的渐进性脑细胞损伤症。如果父亲或母亲患有这种病,下一代就有50%的机会得这种病(除非这个父亲或母亲属于显性纯合子,那么下一代100%都会得这种病)。尽管曾有一些病人在20多岁就开始发病的报道,但这种病一般在30岁以前不会发作。

多指(或多趾)畸形症的症状就是有6个手指或脚趾。现如今我们可以在患者出生时就切除多余的手指或脚趾,这样患者就不会知道他们具有这种性状。亨利八世的一个妻子就有一个多余的手指。在南方的一些家庭,这种性状比较普遍。多余的手指或脚趾没有什么作用,而在买手套时肯定会引起麻烦,更不用说在谋杀案中易被认定出来。

Unit 4

Text A

Microevolution and Macroevolution

Changes in gene frequency that occur within a population without producing a new species are called microevolution. As microevolution continues, a population may become so different that it is no longer able to reproduce with members of other populations. At that point, the population becomes a new species. As the new species continues to evolve, biologists might eventually consider it to be a new genus, order, family, or higher level of classification. Such evolution at the level of species or higher is called macroevolution.

Microevolution can occur very quickly, indeed, it is probably always occurring. For example, in less than half a century after the discovery of antibiotics, many bacteria evolved resistance to them. Resistance to antibiotics evolves when antibiotics are used improperly, allowing the survival of a few bacteria with mutated genes that confer resistance. Natural selection then leads to the evolution of antibiotic-resistant strains. Pesticide-resistant insects and herbicide-resistant weeds are additional examples of rapid microevolution.

Macroevolution occurs over much longer periods and is seldom observed within the human life span. Occasionally, however, scientists do see evidence that new species have recently evolved. There are species of parasitic insects, for example, that are unable to reproduce except in domesticated plants that did not even exist a few centuries ago. The pace of evolution can be quite variable, with long periods in which there is little change being punctuated by relatively brief periods of tens of thousands of years in which most changes occur. This idea that the pace of evolution is not always slow and constant is referred to as punctuated equilibrium. It was first proposed by paleontologists Niles Eldredge and Stephen Jay Gould in 1979, and it is one of many examples of how scientists' views of evolution are continually changing.

Several possible mechanisms exist for rapid evolution. Chromosomal aberrations, such as breakages and rejoining of chromosomal parts, can introduce large changes in genes and the sequences that regulate them. This may lead to changes much larger than that brought about by simple point mutations. Environmental catastrophes can set the stage for rapid evolution as well. It is thought that the extinction of the dinosaurs was triggered by a large comet impact. This rapid loss of the dominant fauna in many ecosystems opened up many new niches for mammals, which at the time were a small group of fairly unimportant creatures. The sudden appearance of many new opportunities led to rapid and widespread speciation, in a process called adaptive radiation.

Other areas of biology are also continually changing under the influence of evolution. For example, as Charles Darwin predicted in The Origin of Species, classification has become more than simply the grouping of organisms into species, genera, families, and so on based on how physically similar they are. Classification now aims to group species according to their evolutionary history. Thus two species that diverged recently from the same ancestor should be in the same genus, whereas species that shared a more distant common ancestor might be in different genera or higher taxonomic levels.

Until the 1980s, evolutionary history, or phylogeny, of organisms could only be inferred from anatomical similarities. Since that time, however, it has been possible to determine phylogeny from comparisons of molecules. Often this molecular phylogeny agrees with the phylogeny based on anatomy. For example, about 99 percent of the sequence of bases in the deoxyribonucleic acid (DNA) of chimpanzees and humans is identical. This finding confirms the conclusion from anatomy that chimpanzees and humans evolved from the same ancestor only a few million years ago. Such agreement between anatomical and molecular phylogeny would not be expected if each species were a totally different creation unrelated to other species, but it makes sense in light of evolution. It is one of many examples of the famous saying by the geneticist Theodosius Dobzhansky (1900-1975): "Nothing in biology makes sense except in the light of evolution."

New Words

microevolution	['maikrəu,i:və'lu:ʃən]	n.	小进化,微进化
macroevolution	[,mækrəuevə'lu:ʃən]	n.	大进化,宏进化
population	[,pɔpju'leiʃən]	n.	种群
species	['spi:ʃiz]	n.	种类,种
reproduce	[,ri:prə'dju:s]	v.	繁殖,再生
evolve	[i'vɔlv]	v.	发展,进展,进化
biologist	[bai'ɔlədʒist]	n.	生物学家
genus	['dʒi:nəs]	n.	属,种,类
order	['ɔ:də]	n.	目

eventually	[iˈventjuəli]	adv.	最后,终于
classification	[ˌklæsifiˈkeiʃən]	n.	分类,分级
evolution	[ˌi:vəˈlu:ʃn]	n.	演变,进化
evidence	[ˈevidəns]	n.	证明,证据
occur	[əˈkə:]	vi.	发生,出现
observe	[əbˈzə:v]	vt.	观察,观测
antibiotic	[ˌæntibaiˈɔtik]	n.	抗生素
		adj.	抗生的
bacteria	[bækˈtiəriə]	n.	细菌
resistance	[riˈzistəns]	n.	抗性
improper	[imˈprɔpə]	adj.	不合适的;错误的;不正常的
survival	[səˈvaivəl]	n.	生存,幸存
mutate	[mju:ˈteit]	v.	变异
strain	[strein]	n.	品系
parasitic	[ˌpærəˈsitik]	adj.	寄生的
punctuate	[ˈpʌŋktjueit]	vt.	不时打断;屡次打断
pace	[peis]	n.	速度
paleontologist	[ˌpæliɔnˈtɔlədʒi]	n.	古生物学者
mechanism	[ˈmekənizəm]	n.	机制
aberration	[ˌæbəˈreiʃən]	n.	失常,畸变
breakage	[ˈbreikidʒ]	n.	破裂
regulate	[ˈregjuleit]	vt.	控制;调整
environmental	[inˌvaiərənˈmentl]	adj.	环境的
catastrophe	[kəˈtæstrəfi]	n.	大灾难,大祸
extinction	[iksˈtiŋkʃən]	n.	消失;灭绝
dinosaur	[ˈdainəsɔ:]	n.	恐龙
trigger	[ˈtrigə]	vt.	引发,引起
comet	[ˈkɔmit]	n.	彗星
impact	[ˈimpækt]	n.	碰撞
niche	[nitʃ]	n.	恰当的处所;小生境
opportunity	[ˌɔpəˈtju:niti]	n.	机会,时机
speciation	[ˌspi:ʃiˈeiʃən]	n.	物种形成
predict	[priˈdikt]	v.	预知,预言,预报
genera	[ˈdʒenərə]	n.	类,属(genus 的复数)
similar	[ˈsimilə]	adj.	相似的,类似的
group	[gru:p]	vt.	分组;分类;归类
ancestor	[ˈænsistə]	n.	物种原形,原种;祖先

taxonomic	[tæksəˈnɔmik]	adj.	分类学的
phylogeny	[faiˈlɔdʒini]	n.	动植物种类进化史,发展史
organism	[ˈɔːgənizəm]	n.	生物体,有机体
infer	[inˈfəː]	v.	推断
anatomical	[ˌænəˈtɔmikəl]	adj.	解剖的,解剖学的
similarity	[ˌsimiˈlæriti]	n.	类似,类似处
anatomy	[əˈnætəmi]	n.	解剖学;剖析
sequence	[ˈsiːkwəns]	n.	次序,顺序
base	[beis]	n.	基
chimpanzee	[tʃimpənˈziː]	n.	黑猩猩
identical	[aiˈdentikəl]	adj.	同一的,同样的
confirm	[kənˈfəːm]	vt.	确定,证实,肯定

Phrases

gene frequency	基因频率
no longer	不再
less than	不到
mutated gene	变异基因
lead to	导致
pesticide-resistant insect	抗杀虫剂昆虫
herbicide-resistant weed	抗除草剂杂草
life span	生命期
domesticated plant	寄主植物
punctuated equilibrium	点平衡
bring about	带来,产生
point mutation	点突变
as well	也
dominant fauna	优势动物群
open up	开辟,打开
adaptive radiation	适应性辐射
The Origin of Species	《物种起源》(达尔文著)
aim to	以……为目标,企图
make sense	有意义
in light of	以……的观点;按照,根据

Notes

[1] Occasionally, however, scientists do see evidence that new species have recently evolved.

本句中,do 表示强调,它强调的是动词 see,意思是"一定,务必"。请看下例:

Please do come to our discussion this Friday.

请这周五一定来参加我们的讨论。

that new species have recently evolved 是 evidence 的同位语。

[2] There are species of parasitic insects, for example, that are unable to reproduce except in domesticated plants that did not even exist a few centuries ago.

本句中,that are unable to reproduce except in domesticated plants that did not even exist a few centuries ago 是一个定语从句,修饰和限定 species of parasitic insects,在该从句中,that did not even exist a few centuries ago 也是一个定语从句,修饰和限定 domesticated plants。

[3] The pace of evolution can be quite variable, with long periods in which there is little change being punctuated by relatively brief periods of tens of thousands of years in which most changes occur.

本句中,in which there is little change being punctuated by relatively brief periods of tens of thousands of years in which most changes occur 是一个介词前置的定语从句,修饰和限定 long periods,在该从句中含有另一个介词前置的定语从句 in which most changes occur,修饰和限定 brief periods of tens of thousands of years。

[4] It is thought that the extinction of the dinosaurs was triggered by a large comet impact.

本句中,It is thought that 是一个句型,意思是"有人认为……,人们认为……"。常用于该句型的动词有:believe,suppose,say,expected,report 等。请看下例:

It is reported that 20 people were injured in the accident.

据报道,在这次交通事故中有 20 人受伤。

Exercises

【EX. 1】 根据课文内容,回答以下问题

1) What is microevolution?

2) What is macroevolution?

3) What is the difference between microevolution and macroevolution?

4) What does natural selection lead to?

5) What is punctuated equilibrium?

6) What can chromosomal aberrations do?

7) What do people think triggered extinction of the dinosaurs?

8) What does classification now aim to do?

9) How many percent of the sequence of bases in the deoxyribonucleic acid (DNA) of chimpanzees and humans is identical?

10) What is the conclusion from anatomy?

【EX. 2】 根据下面的英文解释,写出相应的英文词汇

英 文 解 释	词 汇
Evolution resulting from a succession of relatively small genetic variations that often cause the formation of new subspecies.	
Large-scale evolution occurring over geologic time that results in the formation of new taxonomic groups.	
A fundamental category of taxonomic classification, ranking below a genus or subgenus and consisting of related organisms capable of interbreeding.	
A gradual process in which something changes into a different and usually more complex or better form.	
A deviation in the normal structure or number of chromosomes in an organism.	
The science of the shape and structure of organisms and their parts.	
The capacity of an organism to defend itself against a disease.	
All the organisms that constitute a specific group or occur in a specified habitat.	

英 文 解 释	词 汇
A sudden violent change in the earth's surface; a cataclysm.	
The evolutionary development and history of a species or higher taxonomic grouping of organisms.	
Support or establish the certainty or validity of; verify.	

【EX. 3】 把下列句子翻译为中文

1) Biological evolution is a change in the genetic characteristics of population of organisms over time. For evolution to occur, there must be inherited differences among the individuals in a population.

2) Populations evolve when individuals with certain characters (such as large body size) leave more offspring than other individuals. The inherited characters of the individuals that leave more offspring become more common in the following generation.

3) Adaptations are features of an organism that improve its performance in its environment. Adaptations are products of natural selection, the process in which individuals with particular inherited characters survive and reproduce at a higher rate than other individuals because of those characters.

4) The great diversity of life on Earth has resulted from the repeated splitting of species into two or more species.

5) When one species splits into two, the two species that result share many features because they have evolved from a common ancestor.

6) The evidence that evolution has occurred is overwhelming. One strong line of evidence comes from the fossil record, which allows biologists to reconstruct the history of life on Earth and shows how new species arose from previous species.

7) Natural populations provide clear evidence of evolutionary change.

8) When differences in genes or proteins are examined, species that are thought to be closely related based on the fossil record may be more similar than species thought to be distantly related.

9) Several indirect lines of evidence argue that macroevolution has occurred, including successive changes in homologous structures, developmental patterns, vestigial structures, parallel patterns of evolution, and patterns of distribution.

10) Darwin's theory of evolution has proven controversial among the general public, although the commonly raised objections are without scientific merit.

【EX. 4】 把下列短文翻译为中文

Traits can evolve by natural selection only if they are at least partly heritable. However, individuals may acquire new traits via cultural evolution—that is, by learning them from other individuals. Cultural evolution is most highly developed in humans, whose language and remarkable learning abilities enable new innovations to spread and be adopted at rapid rates. But the only requirement for traits to evolve via cultural evolution is that individuals have the ability to learn them. Birds, for example, copy the songs of other individuals, resulting in the evolution of song "dialects." Many behaviors of the apes are transmitted via learning. In one study, investigators compared the behavior of four orang-outang populations on the island of Borneo and two on Sumatra. The investigators identified 24 behaviors that are restricted to a single population. These behaviors are not correlated with any differences in the environments in which the populations live. Ten of the behaviors are specialized feeding techniques, including tool use. Six are alternative forms of social signals, such as kiss-squeaks. Thus, orangutan populations develop cultural distinctions as individuals copy the behavior of other individuals.

Text B

Darwinian Evolution

Modern biology is based on several unifying themes, such as the cell theory, genetics and inheritance, Francis Crick's central dogma of information flow, and Darwin and Wallace's theory of evolution by natural selection. In this text we will examine these themes and the nature of science.

Charles Darwin, former divinity student and former medical student, secured (through the intercession of his geology professor) an unpaid position as ship's naturalist on the British exploratory vessel H. M. S. Beagle. The voyage would provide Darwin a unique opportunity to study adaptation and gather a great deal of proof he would later incorporate into his theory of evolution. On his return to England in 1836, Darwin began (with the assistance of numerous specialists) to catalog his collections and ponder the seeming "fit" of organisms to their mode of existence. He eventually settled on four main points of a radical new hypothesis:

1. Adaptation: all organisms adapt to their environments.

2. Variation: all organisms are variable in their traits.

3. Over-reproduction: all organisms tend to reproduce beyond their environment's capacity to support them (this is based on the work of Thomas Malthus, who studied how populations of organisms tended to grow geometrically until they encountered a limit on their population size).

4. Since not all organisms are equally well adapted to their environment, some will survive and reproduce better than others — this is known as natural selection. Sometimes this is also referred to as "survival of the fittest". In reality this merely deals with the reproductive success of the organisms, not solely their relative strength or speed.

Unlike the upper-class Darwin, Alfred Russel Wallace (1823-1913) came from a different social class. Wallace spent many years in South America, publishing salvaged notes in Travels on the Amazon and Rio Negro in 1853. In 1854, Wallace left England to study the natural history of Indonesia, where he contracted malaria. During a fever Wallace managed to write down his ideas on natural selection.

In 1858, Darwin received a letter from Wallace, in which Darwin's as-yet-unpublished theory of evolution and adaptation was precisely detailed. Darwin arranged for Wallace's letter to be read at a scientific meeting, along with a synopsis of his own ideas. To be correct, we need to mention that both Darwin and Wallace developed the theory, although Darwin's major work was not published until 1859 (On the Origin of Species by Means of Natural Selection). While there have been some changes to the theory since 1859, most notably the incorporation of genetics and DNA into what is termed the "Modern Synthesis" during the 1940s, most scientists today accept evolution as the guiding theory on which modern biology is based.

Recent revisions of biology curricula stressed the need for underlying themes. Evolution serves as such a universal theme. Evolutionary theory and the cell theory provide us with a basis for the interrelation of all living things. We also utilize Linneus' hierarchical classification system, adopting (generally) five kingdoms of living organisms. The five kingdoms are: Monera, Protista, Fungi, Plantae and Animalia.

New Words

unifying	[ˈjuːnifaiiŋ]	adj. 统一的，一体的
theme	[θiːm]	n. 题目，主题
genetics	[dʒiˈnetiks]	n. 遗传学；发生学

divinity	[di'viniti]	n. 神,神学,神性;上帝
secure	[si'kjuə]	vt. 赢得,获得,取得
		adj. 安全的,可靠的
intercession	[intə'seʃ(ə)n]	n. 说情,调解
geology	[dʒi'ɔlədʒi]	n. 地质学;地质概况
naturalist	['nætʃərəlist]	n. 博物学家;自然主义者
exploratory	[iks'plɔrətəri]	adj. 探险的,探测的
vessel	['vesl]	n. 船;容器,器皿;脉管
unique	[ju'ni:k]	adj. 惟一的,独一无二的,独特的;少见的
opportunity	[ɔpə'tju:niti]	n. 机会,时机
adaptation	[ədæp'teiʃ(ə)n]	n. 适应
specialist	['speʃəlist]	n. 专家,行家
catalog	['kætəlɔg]	n. 目录,目录册
		vt. 编目录
collection	[kə'lekʃən]	n. 收藏品;收集物;收集,收取
ponder	['pɔndə]	vt. 沉思,考虑
seeming	['si:miŋ]	adj. 表面上的,外观上的;似乎是的
		n. 外观
eventually	[i'ventjuəli]	adv. 最后,终于
radical	['rædik(ə)l]	adj. 基本的;根本的;彻底的;完全的;激进的;极端的
hypothesis	[hai'pɔθisis]	n. 假设;前提
variation	[ˌveəri'eiʃən]	n. 变异,变种,变更,变化
variable	['veəriəbl]	adj. 可变的,不定的,易变的
trait	[treit]	n. 显著的特点,特性
reproduce	[ˌri:prə'dju:s]	vt. & vi. 繁殖,再生,复制
population	[ˌpɔpju'leiʃən]	n. 人口;(动物的)种群
geometrically	[dʒiə'metrikəli]	adv. 用几何学,几何学上
encounter	[in'kauntə]	vt. 遭遇,遇到,相遇
survive	[sə'vaiv]	vt. & vi. 幸免于,幸存,生还
merely	['miəli]	adv. 仅仅,只,不过
reproductive	[ˌri:prə'dʌktiv]	adj. 生殖的,再生的,复制的
success	[sək'ses]	n. 成功,成就,胜利
solely	['səuli]	adv. 只是;独自地,单独地
publish	['pʌbliʃ]	vt. 出版,刊印,公布,发表
Indonesia	[ˌində'ni:ziə]	n. 印度尼西亚
contract	['kɔntrækt]	v. 感染;订约

		n.	合同,契约,婚约
malaria	[məˈlɛəriə]	n.	疟疾,瘴气
synopsis	[siˈnɔpsis]	n.	大意;要略;纲要;大纲
hierarchical	[ˌhaiəˈrɑːkikəl]	adj.	分等级的
classification	[ˌklæsifiˈkeiʃən]	n.	分类法;分类,分级
kingdom	[ˈkiŋdəm]	n.	领域;范围;王国
Monera	[ˈmɔnərə]	n.	原核生物界
protista	[ˈprəutistə]	n.	原生生物
Fungi	[ˈfʌndʒai, ˈfʌŋgai]	n.	真菌类
Plantae	[ˈplænti:]	n.	植物界
Animalia	[ˌæniˈmeiliə]	n.	动物界,动物类

Phrases

Darwinian Evolution	达尔文的进化论
be based on	基于……,以……为基础
natural selection	自然选择
information flow	信息流
a great deal of	许多,大量
incorporate... into...	把……合并到……,把……并入……
mode of existence	存在模式
settle on	停留,落在
adapt to	适应,习惯于
survival of the fittest	适者生存
in reality	实际上,事实上
deal with	处理,对付

Exercises

【EX. 5】 根据课文内容,回答以下问题

1) What is modern biology based on?

2) What did Charles Darwin study before he had the chance to study adaptation?

3) What were the four main points of a radical new hypothesis Charles Darwin settled on?

4) When did Darwin begin (with the assistance of numerous specialists) to catalog his collections and ponder the seeming "fit" of organisms to their mode of existence?

5) What did he eventually settled on?

6) What did Darwin receive from Wallace in 1858? What was in it?

7) What do evolutionary theory and the cell theory provide us with?

8) What are the five kingdoms of living organisms?

Reading Material

Text	Notes
Darwin's Critics In the century since he proposed it, Darwin's theory of evolution by natural selection has become nearly universally accepted by biologists, but has proven controversial[1] among the general public. Darwin's critics raise seven principal objections to teaching evolution: 1. Evolution is not solidly demonstrated. "Evolution is just a theory," Darwin's critics point out, as if theory meant lack of knowledge, some kind of guess. Scientists, however, use the word theory in a very different sense than the general public does. Theories are the solid ground of science, that of which we are most certain. Few of us doubt the theory of gravity because it is "just a theory." 2. There are no fossil intermediates. "No one ever saw a fin on the way to becoming a leg," critics claim, pointing to the many gaps in the fossil record in Darwin's day. Since then, however, most fossil intermediates in vertebrate evolution have indeed been found. A clear line of fossils now traces the transition between whales[2] and hoofed[3] mammals, between reptiles and mammals, between dinosaurs and birds, between apes and humans. The fossil evidence of evolution between major forms is compelling[4].	[1]引起争论的，有争议的 [2]鲸 [3]有蹄的 [4]强制的，强迫的

Text	Notes
3. The intelligent design argument. "The organs of living creatures are too complex for a random process to have produced—the existence of a clock is evidence of the existence of a clockmaker." Biologists do not agree. The intermediates in the evolution of the mammalian ear can be seen in fossils, and many intermediate "eyes" are known in various invertebrates. These intermediate forms arose because they have value—being able to detect light a little is better than not being able to detect it at all. Complex structures like eyes evolved as a progression[5] of slight improvements. 4. Evolution violates[6] the Second Law of Thermodynamics. "A jumble of soda cans doesn't by itself jump neatly into a stack—things become more disorganized due to random events, not more organized." Biologists point out that this argument ignores what the second law really says: disorder increases in a closed system, which the earth most certainly is not. Energy continually enters the biosphere[7] from the sun, fueling life and all the processes that organize it. 5. Proteins are too improbable. "Hemoglobin has 141 amino acids. The probability that the first one would be leucine is $1/20$, and that all 141 would be the ones they are by chance is $(1/20)^{141}$, an impossibly rare event." This is statistical foolishness—you cannot use probability[8] to argue backwards. The probability that a student in a classroom has a particular birthday is $1/365$; arguing this way, the probability that everyone in a class of 50 would have the birthdays they do is $(1/365)^{50}$, and yet there the class sits. 6. Natural selection does not imply evolution. "No scientist has come up with an experiment where fish evolve into frogs and leap away from predators[9]." Is microevolution (evolution within a species) the mechanism that has produced macroevolution (evolution among species)? Most biologists that have studied the problem think so. Some kinds of animals produced by artificial selection are remarkably distinctive, such as Chihuahuas[10], dachshunds[11], and greyhounds[12]. While all dogs are in fact the same species and can interbreed, laboratory selection experiments easily create forms that cannot interbreed and thus would in nature	[5]进步,发展 [6]违犯;违反 [7]生物圈 [8]可能性,或然性,概率 [9]食肉动物 [10]吉娃娃(一种产于墨西哥的狗) [11]达克斯猎犬 [12]灰狗

Text	Notes
be considered different species. Thus, production of radically different forms has indeed been observed, repeatedly. To object that evolution still does not explain really major differences, like between fish and amphibians, simply takes us back to point 2. These changes take millions of years, and are seen clearly in the fossil record.	
7. The irreducible[13] complexity argument. The intricate molecular machinery of the cell cannot be explained by evolution from simpler stages. Because each part of a complex cellular process like blood clotting is essential to the overall process, how can natural selection fashion any one part? What's wrong with this argument is that each part of a complex molecular machine evolves as part of the system. Natural selection can act on a complex system because at every stage of its evolution, the system functions. Parts that improve function are added, and, because of later changes, become essential. The mammalian blood clotting system, for example, has evolved from much simpler systems. The core clotting system evolved at the dawn of the vertebrates 600 million years ago, and is found today in lampreys[14], the most primitive fish. One hundred million years later, as vertebrates evolved, proteins were added to the clotting system making it sensitive to substances released from damaged tissues. Fifty million years later, a third component was added, triggering clotting by contact with the jagged surfaces produced by injury. At each stage as the clotting system evolved to become more complex, its overall performance came to depend on the added elements. Thus, blood clotting has become "irreducibly complex"—as the result of Darwinian evolution.	[13] 不能复归的,不能削减的 [14] 七鳃鳗

Text A 参考译文

小进化和大进化

　　小进化是指出现在不产生新物种的一个种群内基因频率的改变。小进化积累后,一个种群变得与其他种群明显不同,这个种群内的个体不再能与其他种群的个体繁殖后代。那时候,这个种群就成为一个新的物种。由于这种新物种连续进化,生物学家可能最终认为它属于一种新的属、目、科或分类学上的更高层次中的物种。这种在种或更高层次上的进化叫大进化。

小进化进程相当迅速;事实上,它可能无处不在。例如,在抗生素发现后的不到半个世纪里,许多细菌进化了对它们的抗性。由于抗生素使用不当,抗生素的抗性进化了,使一些带有抗性变异基因的细菌可以存活。自然选择导致抗生素抗性品系的进化。抗杀虫剂昆虫和抗除草剂杂草也是快速小进化的例子。

大进化发生在更长的周期,在人类的生命期内很少被发现。然而,有时候科学家真的看到了刚进化来的新特种的证据。例如,有一些寄生性昆虫的种类,只有在那些几世纪前已经不再存在的寄主植物中才能繁殖。进化的速度可以有很大的不同,在漫长的岁月中几乎没有什么变化,大多数变化是在其间相对短的数万年间发生的。这种进化的速度不总是慢的和持续的观点被称之为点平衡。这是古生物学者 Niles Eldredge 和 Stephen Jay Gould 于 1979 年首次提出的,科学家对进化的观点是不断变化的,这仅是许多例子中的一个。

关于快速进化存在几个可能的机制。染色体畸变,如染色体片段的破裂和重接可以使调节它们的基因和序列发生大变化。这可能导致比简单的点突变大得多的变化。环境的突变也可以引起快速突变。例如,恐龙的灭绝被认为是一颗大彗星碰撞地球而触发的。在许多生态系统中,优势动物群的快速消失为许多哺乳动物让出了生态位,在那个时候,哺乳动物是相对不重要的生物小群体。突然间许多新机遇的出现导致了快速而广泛的物种形成,这个过程叫做适应性辐射。

生物学的其他领域也受进化论的影响而不断地发生变化。例如,查尔斯·达尔文在《物种起源》中预言,分类学已不仅仅根据生物体的相似程度简单地将有机体分为种、属、科等。现代分类学试图依据物种的进化史进行分类。这样两个刚从同一祖先物种分化的物种应该是同一属生物,而共同祖先较远的物种可能是不同的属的生物或属于分类学上更高层次的分类单元中不同的生物。

在 20 世纪 80 年代之前,生物的进化历史或发展史只能通过解剖学的相似性来推断。然而,从那时起,通过分子比较确定进化史有了可能。通常这种分子进化史与基于解剖学的进化史是一致的。例如,黑猩猩和人类的 DNA 中约 99% 的碱基序列是相同的。这个发现证实了解剖学的结论,即黑猩猩和人类是几百万年前同一祖先进化而来的。如果一个物种与其他物种是完全不同的生物,这种解剖学进化史与分子进化史一致性将不存在,但它对进化是有意义的。正如遗传学家 Theodosius Dobzhansky(1900—1975)所说的那样:"除非是放在进化论的背景下,不然生物学就失去了一个整体性的意义。"这只是许多著名的讲话中的一个例子。

Text B 参考译文

达尔文进化论

现代生物学是以几个统一的理论为基础的,如细胞理论、基因和遗传、弗朗西斯·克里克的信息传递的中心法则及达尔文和华莱士的自然选择的进化论。在本单元中我们将就这些理论和科学的本质做一分析。

　　查尔斯·达尔文起先是一名神学学生和医科学生,他(通过他的地质学教授的说情)谋得了一份没有薪水的职位,在英国探险船 H. M. S. Beagle 号上做一名博物学家。这次航海旅行给达尔文提供了独一无二的机会,来研究生物适应性以及收集大量的证据,后来这些证据成为他的进化论的一部分。1836年回到英国之后,达尔文开始(在众多专家的协助之下)对收集的内容进行归类编目,并仔细考虑了生物体的生存方式对环境的适应性。他创立了由4个要点组成的一个新假说。

　　1. 适应性:所有的生物体都能适应它们的生存环境。

　　2. 变异性:所有的生物体的性状都是可变异的。

　　3. 过度繁殖:所有生物的繁殖都倾向于超过所处的环境的承受能力(这一理论基于 Thomas Malthus 的工作,他研究了生物种群是怎样倾向于以几何级方式进行增长,直到面对种群数量极限的挑战)。

　　4. 由于并不是所有的生物都能很好地适应它们的生存环境,那么就会有一些生物的生存和繁殖能力要优于其他的生物,这就是我们所知的自然选择。有时,也可称之为"适者生存"。事实上,这里所说的繁殖能力是指生物体的繁殖成功率,而不是指繁殖的相对强度或速度。

　　与来自上流社会的达尔文不同,阿尔佛雷德·拉塞尔·华莱士(1823—1913)来自于另一社会阶层。华莱士在南美洲待了多年,1853年出版了其在亚马逊河和里约热内卢旅行时的救援记录。1854年,华莱士离开英国,来到印度尼西亚研究当地的自然史,在那里他染上了疟疾,发烧时华莱士写下了他对自然选择的一些看法。

　　1858年,达尔文收到一封来自华莱士的信,在信中达尔文当时尚未发表的进化和适应理论得到准确、详细的说明。在一次学术会议上,达尔文安排阅读华莱士的信,信中内容随同基于自己观点的假说一起被发布。虽然达尔文的主要研究成果直到1859才被发表(涉及通过自然选择的物种起源),但准确地说,这一理论是达尔文和华莱士两人共同发展的。同时,自1859年以来,这个理论已经有了一些变化,其中最重要的是在20世纪40年代补充了遗传学和DNA理论,被定义为"现代假说"的内容。今天的大多数科学家接受了进化论,把它作为当代生物学依据的指导性理论。

　　最新修订版的生物学课程强调了对基本理论的需要。进化就适合作为一个普遍适用的理论。进化论和细胞理论为我们提供了所有生物相互关系的基础。我们也利用林奈的级系分类系统,(通常)沿用生物的五界系统:原核生物类、原生生物类、真菌类、植物类、动物类。

Unit 5

Text A

Plants and Their Structure

A plant has two organ systems: 1) the shoot system, and 2) the root system. The shoot system is above ground and includes the organs such as leaves, buds, stems, flowers (if the plant has any), and fruits (if the plant has any). The root system includes those parts of the plant below ground, such as the roots, tubers, and rhizomes.

Plant cells are formed at meristems, and then develop into cell types which are grouped into tissues. Plants have only three tissue types: 1) Dermal; 2) Ground; and 3) Vascular. Dermal tissue covers the outer surface of herbaceous plants. Dermal tissue is composed of epidermal cells, closely packed cells that secrete a waxy cuticle that aids in the prevention of water loss. The ground tissue comprises the bulk of the primary plant body. Parenchyma, collenchyma, and sclerenchyma cells are common in the ground tissue. Vascular tissue transports food, water, hormones and minerals within the plant. Vascular tissue includes xylem, phloem, parenchyma, and cambium cells.

Plant cell types rise by mitosis from a meristem. A meristem may be defined as a region of localized mitosis. Meristems may be at the tip of the shoot or root (a type known as the apical meristem) or lateral, occurring in cylinders extending nearly the length of the plant. A cambium is a lateral meristem that produces (usually) secondary growth. Secondary growth produces both wood and cork (although from separate secondary meristems).

Parenchyma

A generalized plant cell type, parenchyma cells are alive at maturity. They function in storage, photosynthesis, and as the bulk of ground and vascular tissues. Palisade parenchyma cells are elogated cells located in many leaves just below the epidermal tissue. Spongy mesophyll cells occur below the one or two layers of palisade cells. Ray parenchyma cells occur in wood rays, the structures that transport materials laterally within a woody stem. Parenchyma cells also occur within the xylem and phloem of vascular bundles. The largest parenchyma cells occur in the pith region, often, as in corn (*Zea*) stems, being larger than the vascular bundles. In many prepared slides they stain green.

Collenchyma

Collenchyma cells support the plant. These cells are characterized by thickening of the wall, they are alive at maturity. They tend to occur as part of vascular bundles or on the corners of angular stems. In many prepared slides they stain red.

Sclerenchyma

Sclerenchyma cells support the plant. They often occur as bundle cap fibers. Sclerenchyma cells are characterized by thickening in their secondary walls. They are dead at maturity. They, like collenchyma, stain red in many commonly used prepared slides. A common type of schlerenchyma cell is the fiber.

Xylem

Xylem is a term applied to woody (lignin-impregnated) walls of certain cells of plants. Xylem cells tend to conduct water and minerals from roots to leaves. While parenchyma cells do occur within what is commonly termed the "xylem", the more identifiable cells, tracheids and vessel elements tend to stain red with Safranin-O. Tracheids are the more primitive of the two cell types, occurring in the earliest vascular plants. Tracheids are long and tapered, with angled end plates that connect cell to cell. Vessel elements are shorter, much wider, and lack end plates. They occur only in angiosperms, the most recently evolved large group of plants.

Tracheids, longer, and narrower than most vessels, appear first in the fossil record. Vessels occur later. Tracheids have obliquely-angled endwalls cut across by bars. The evolutionary trend in vessels is for shorter cells, with no bars on the endwalls.

Phloem

Phloem cells conduct food from leaves to rest of the plant. They are alive at maturity and tend to stain green (with the stain fast green). Phloem cells are usually located outside the xylem. The two most common cells in the phloem are the companion cells and sieve cells. Companion cells retain their nucleus and control the adjacent sieve cells. Dissolved food, as sucrose, flows through the sieve cells.

Epidermal Cells

1. Epidermis

The epidermal tissue functions in prevention of water loss and acts as a barrier to fungi and other invaders. Thus, epidermal cells are closely packed, with little intercellular space. To further cut down on water loss, many plants have a waxy cuticle layer deposited on top of the epidermal cells.

2. Guard Cells

To facilitate gas exchange between the inner parts of leaves, stems, and fruits, plants have a series of openings known as stomata (singular stoma). Obviously these openings would allow gas exchange, but at a cost of water loss. Guard cells are bean-shaped cells covering the stomata opening. They regulate exchange of water vapor, oxygen and carbon dioxide through the stoma.

New Words

tuber	['tju:bə]	n.	块茎
rhizome	['raizəum]	n.	根茎;地下茎
meristem	['meristem]	n.	分裂组织,分生组织
dermal	['də:məl]	adj.	皮肤的,真皮的
vascular	['væskjulə]	adj.	维管的;脉管的;导管的;血管的
herbaceous	[hə:'beiʃəs]	adj.	草本的,似绿叶的
epidermal	[ˌepi'də:məl]	adj.	表皮的,外皮的
cuticle	[ˌkju:tikl]	n.	表皮
parenchyma	[pə'reŋkimə]	n.	薄壁组织;软组织
collenchyma	[kə'leŋkimə]	n.	厚角组织
sclerenchyma	[skliə'reŋkimə]	n.	厚壁组织
hormone	['hɔ:məun]	n.	荷尔蒙,激素
xylem	['zailem]	n.	木质组织,木质部
phloem	['fləuem]	n.	韧皮部
cambium	['kæmbiəm]	n.	形成层,新生组织
lateral	['lætərəl]	adj.	横的;侧面的
tracheid	['treikiid]	n.	管胞
tapered	['teipəd]	adj.	锥形的
angiosperm	['ændʒiəuˌspə:m]	n.	被子植物
endwall	['endwɔ:l]	n.	端壁
bar	[bɑ:]	n.	棒
epidermis	[ˌepi'də:mis]	n.	表皮,上皮
barrier	['bæriə]	n.	障碍,阻挡;障碍物
stomata	['stəumətə]	n.	口,气孔(stoma 的复数形式)

Phrases

organ system	器官系统
shoot system	茎系统
root system	根系统
palisade parenchyma cells	栅栏薄壁组织细胞
spongy mesophyll cell	海绵叶肉细胞
ray parenchyma cell	射线薄壁组织细胞
vascular bundle	维管束
vessel element	导管分子
end plate	端板
obliquely-angled endwall	斜角端壁
companion cell	伴胞
sieve cell	筛胞
epidermal cell	表皮细胞
epidermal tissue	表皮组织
guard cell	保卫细胞

Notes

[1] Dermal tissue is composed of epidermal cells, closely packed cells that secrete a waxy cuticle that aids in the prevention of water loss.

本句中,closely packed cells that secrete a waxy cuticle that aids in the prevention of water loss 是一个名词短语,修饰和限定 epidermal cells。在该短语中,that secrete a waxy cuticle 和 that aids in the prevention of water loss 是定语从句,分别修饰和限定 closely packed cells 和 a waxy cuticle。

[2] Palisade parenchyma cells are elogated cells located in many leaves just below the epidermal tissue.

本句中,located in many leaves just below the epidermal tissue 是一个过去分词短语,作定语,修饰和限定 elogated cells。该短语可扩展为一个定语从句:which are located in many leaves just below the epidermal tissue。

[3] While parenchyma cells do occur within what is commonly termed the "xylem", tracheids and vessel elements tend to stain red with Safranin-O.

本句中,While parenchyma cells do occur within what is commonly termed the "xylem"是一个让步状语从句,修饰主句的谓语 tend to。在该从句中,do 表示强调,意思是"确实,的确"。

Exercises

【EX. 1】 根据课文内容,回答以下问题

1) How many organ systems does a plant have? What are they?

2) What does the shoot system include? What about the root system?

3) How many tissue types do plants have? What are they?

4) How may a meristem be defined?

5) What is a cambium?

6) What is the difference between collenchyma cells and sclerenchyma cells?

7) What is xylem?

8) What are tracheids and vessel elements?

9) What is the function of the epidermal tissue?

10) What do guard cells do?

【EX. 2】 根据下面的英文解释,写出相应的英文词汇

英　文　解　释	词　汇
The supporting and water-conducting tissue of vascular plants, consisting primarily of tracheids and vessels; woody tissue.	
The primary tissue of higher plants, composed of thin-walled cells and forming the greater part of leaves, roots, the pulp of fruit, and the pith of stems.	
The food-conducting tissue of vascular plants, consisting of sieve tubes, fibers, parenchyma, and sclereids.	
A cell in the xylem of vascular plants.	

续表

英 文 解 释	词 汇
A horizontal, usually underground stem that often sends out roots and shoots from its nodes.	
A swollen, fleshy, usually underground stem, such as the potato, bearing buds from which new plant shoots arise.	
A supportive plant tissue that consists of thick-walled, usually lignified cells.	
The outer, protective, nonvascular layer of the skin of vertebrates, covering the dermis, cuticle.	
A plant whose ovules are enclosed in an ovary; a flowering plant.	
Relating to or characteristic of an herb as distinguished from a woody plant.	

【EX. 3】 把下列句子翻译为中文

1) Meristems are actively dividing, embryonic tissues responsible for both primary and secondary growth.

2) Plant basic life cycle involved in alternation of generations can be summarized as zygote- sporophyte- meiosis-spores - gametophyte-gametes (eggs and sperm)- syngamy-zygote. Such a cycle is characteristic of all plants.

3) Plants grow from the division of meristematic tissue. Primary growth results from cell division at the apical meristem at the tip of the plant, making the shoot longer.

4) The growth of leaves is determinate, like that of flowers; the growth of stems and roots is indeterminate. In determinate growth, the meristematic cells eventually cease to divide; in indeterminate growth, they retain the capacity to divide indefinitely.

5) Xylem conducts water and dissolved minerals from the roots to the shoots and the leaves. Phloem carries organic materials from one part of the plant to another.

6) Transport systems, external barriers, and a branching root system develop from the primary root as it matures.

7) Some plants have modified roots that carry out photosynthesis, gather oxygen, parasitize other plants, store food or water, or support the stem.

8) In angiosperms, both male and female structures often occur together in the same individual flower. These reproductive structures are not a permanent part of the adult individual and the germ line is not set aside early in development.

9) Bees are the most frequent and characteristic pollinators of flowers. Insects are often attracted by the odor of flowers. Bird-pollinated flowers are characteristically odorless and red, with the nectar not readily accessed by insects.

10) In double fertilization, angiosperms utilize two sperm cells. One fertilizes the egg, while the other helps form a substance called endosperm that nourishes the embryo.

【EX. 4】 把下列短文翻译为中文

The major source of plant nutrition is the fixation of atmospheric CO_2 into simple sugar using the energy of the sun. CO_2 enters through the stomata. O_2 is a product of photosynthesis and atmospheric component that also moves through the stomata. It is used in cellular respiration to release energy from the chemical bonds in the sugar to support growth and maintenance in the plant. However, CO_2 and light energy are not sufficient for the synthesis of all the molecules a plant needs. Plants require a number of inorganic nutrients. Some of these are macronutrients, which the plants need in relatively large amounts, and others are micronutrients, which are required in trace amounts. There are nine macronutrients: carbon, hydrogen, and oxygen—the three elements found in all organic compounds—as well as nitrogen (essential for amino acids), potassium, calcium, phosphorus, magnesium (the center of the chlorophyll molecule), and sulfur. Each of these nutrients approaches or, as in the case with carbon, may greatly exceed 1% of the dry weight of a healthy plant. The seven micronutrient elements——iron, chlorine, copper, manganese, zinc, molybdenum, and boron——constitute from less than one to several hundred parts per million in most plants.

Text B

Flowers

Flowers are collections of reproductive and sterile tissue arranged in a tight whorled array having very short internodes. Sterile parts of flowers are the sepals and petals. When these are similar in size and shape, they are termed tepals. Reproductive parts of the flower are the stamen (male, collectively termed the androecium) and carpel (often the carpel is referred to as the pistil, the female parts collectively termed the gynoecium).

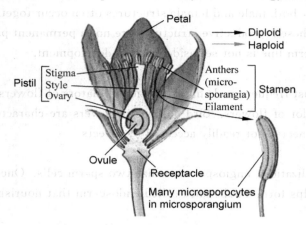

The structure of a flowering plant

Androecium

The individual units of the androecium are the stamens, which consist of a filament which supports the anther. The anther contains four microsporangia within which microspores (pollen) are produced by meiosis.

Stamens are thought to represent modified sporophylls (leaves with sporangia on their upper surface). Examinations by James E. Canright in the 1950s suggested an evolutionary series from primitive angiosperms which have leafish stamens to others with "normal" stamens (*Lilium*).

Pollen

Pollen grains (from the greek *palynos* for dust or pollen) contain the male gametophyte (microgametophyte) phase of the plant. Pollen grains are produced by meiosis of microspore mother cells that are located along the inner edge of the anther sacs (microsporangia). The outer part of the pollen is the exine, which is composed of a complex polysaccharide, sporopollenin. Inside the pollen are two (or, at most, three) cells that comprise the male gametophyte. The tube cell (also referred to as the tube nucleus) develops into the pollen tube. The germ cell divides by mitosis to produce two sperm cells. Division of the germ cell can occur before or after pollination.

Gynoecium

The gynoecium consists of the stigma, style, and ovary containing one or more ovules. These three structures are often termed a pistil or carpel. In many plants, the pistils will fuse for all or part of their length.

Like the stamen, the carpel is thought to be a modified leaf. Work by I. W. Bailey and his students pointed to an evolutionary sequence from primitive angiosperms (like *Drimys*) to "normal" carpels like those of *Lilium*.

The Stigma and Style

The stigma functions as a receptive surface on which pollen lands and germinates its pollen tube. Corn silk is part stigma, part style. The style serves to move the stigma some distance from the ovary. This distance is species specific.

The Ovary

The ovary contains one or more ovules, which in turn contain one female gametophyte, also referred to in angiosperms as the embryo sac. Some plants, such as cherry, have only a single ovary which produces two ovules. Only one ovule will develop into a seed.

The Gametophytes

The male gametophyte develops inside the pollen grain. The female gametophyte develops inside the ovule. In flowering plants, gametophyte phases are reduced to a few cells dependant for their nutrition in the sporophyte phase. This is the reverse of the pattern seen in the nonvascular plant groups liverworts, mosses, and hornworts.

Angiosperm male gametophytes have two haploid nuclei (the germ nucleus and tube nucleus) contained within the exine of the pollen grain (or microspore).

Female gametophytes of flowering plants develop within the ovule (megaspore) contained within an ovary at the base of the pistil of the flower. There are usually eight (haploid) cells in the female gametophyte: a) one egg, two synergids flanking the egg (located at the micropyle end of the embryo sac); b) two polar nuclei in the center of the embryo sac; and three antipodal cells (at the opposite end of the embryo sac from the egg).

Double Fertilization

The process of pollination being accomplished, the pollen tube grows through the stigma and style toward the ovules in the ovary. The germ cell in the pollen grain divides and releases two sperm cells which move down the pollen tube. Once the tip of the tube reaches the micropyle end of the embryo sac, the tube grows through into the embryo sac through one of the synergids which flank the egg. One sperm cell fuses with the egg, producing the zygote which will later develop into the next-generation sporophyte. The second sperm fuses with the two polar bodies located in the center of the sac, producing the nutritive triploid endosperm tissue that will provide energy for the embryo's growth and development.

Fruit

The ovary wall, after fertilization has occurred, develops into a fruit. Fruits may be fleshy, hard, multiple or single.

Vegetative Propagation

Many plants also have an asexual method of reproduction. Often some species, such as many orchids, are more frequently propagated vegetatively than via seeds. Tubers are fleshy underground stems, as in the Irish potato. Leaflets are sections of leaf that will develop roots and drop off the plant, effectively cloning the plant. Runners are shoots running along or over the surface of the ground that will sprout a plantlet, which upon settling to the ground develop into a new independent plant.

Plant Nutrition

Unlike animals (which obtain their food from what they eat) plants obtain their nutrition from the soil and atmosphere. Using sunlight as an energy source, plants are capable of making all the organic macromolecules they need by modifications of the sugars they form by photosynthesis. However, plants must take up various minerals through their root systems for use.

New Words

sterile	['sterail]	adj.	不育的,不结果的
whorled	[hwə:ld]	adj.	有涡旋的,有螺纹的
internode	['intəˌnəud]	n.	节间,结间部
sepal	['sepəl]	n.	萼片
petal	['petəl]	n.	花瓣
tepal	['tepəl]	n.	被片
stamen	['steimen]	n.	雄蕊,雄性花蕊
androecium	[æn'dri:ʃiəm]	n.	雄蕊
carpel	['kɑ:pel]	n.	心皮
pistil	['pistil]	n.	雌蕊
gynoecium	[dʒi'ni:siəm]	n.	雌蕊,雌蕊群

filament	[ˈfiləmənt]	n.	细丝，花丝
anther	[ˈænθə]	n.	花药
pollen	[ˈpɔlin]	n.	花粉
angiosperm	[ˈændʒiəuˌspə:m]	n.	被子植物
microsporangia	[ˌmaikrəuspəˈrændʒiə]	n.	小孢子囊（microsporangium 的复数）
sporophyll	[ˈspɔ:rəˌfil]	n.	孢子叶
lilium	[ˈliliəm]	n.	百合属植物
gametophyte	[gəˈmi:təˌfait]	n.	配偶体
sac	[sæk]	n.	囊；液囊
exine	[ˈeksi:n]	n.	花粉粒外壁
sporopollenin	[ˌspɔ:rəˈpɔlənin]	n.	孢子花粉素
pollination	[pɔliˈneiʃən]	n.	授粉
stigma	[ˈstigmə]	n.	柱头
style	[stail]	n.	花柱
ovary	[ˈəuvəri]	n.	子房
ovule	[ˈəuvju:l]	n.	胚珠
liverwort	[ˈlivəˌwə:t]	n.	地钱
moss	[mɔs]	n.	苔，藓
hornwort	[ˈhɔ:nwə:t]	n.	金鱼藻，角苔
haploid	[ˈhæplɔid]	adj.	单元体的，单倍体的；单一的
microspore	[ˈmaikrəuspɔ:]	n.	小孢子；花粉粒
megaspore	[ˈmegəspɔ:]	n.	大孢子
synergid	[siˈnə:dʒid]	n.	助细胞
micropyle	[ˈmaikrəupail]	n.	珠孔；卵孔
antipodal	[ænˈtipədl]	adj.	对跖的，正相反的
sporophyte	[ˈspɔ:rəˌfait]	n.	孢子体，孢子形成体
endosperm	[ˈendəuspə:m]	n.	胚乳
asexual	[æˈseksuəl]	adj.	无性的，无性生殖的
propagate	[ˈprɔpəgeit]	vt.	繁殖
orchid	[ˈɔ:kid]	n.	兰，兰花
clone	[kləun]	v.	无性繁殖，克隆
		n.	无性系，无性繁殖，克隆
runner	[ˈrʌnə]	n.	长匍茎
sprout	[spraut]	v.	长出，冒出；发芽，萌芽
plantlet	[ˈplɑ:ntlit]	n.	小植物，苗
mineral	[ˈminərəl]	n.	矿物，矿物质

Phrases

pollen grain	花粉粒
microspore mother cell	小孢子母细胞
complex polysaccharide	复合多糖
tube cell	管细胞
pollen tube	花粉管
germ cell	生殖细胞,精子,卵子,受精卵
sperm cell	精细胞
embryo sac	胚囊
nonvascular plant groups	非维管植物组
germ nucleus	精核
tube nucleus	管核
double fertilization	双授精

Exercises

【EX. 5】 根据课文内容,回答以下问题

1) What are flowers?

2) What are sterile parts and reproductive parts of flowers?

3) What are the individual units of the androecium? What do they consist of?

4) What does the anther contain?

5) What does the gynoecium consist of?

6) What does the stigma function as?

7) Where does the male gametophyte develop? What about the female gametophyte?

8) What does one sperm cell do? What about the second?

9) What may fruits be?

10) What are runners?

Reading Material

Text	Notes
Plant Balanced Diet Carbon, Hydrogen, and Oxygen are considered the essential elements. Nitrogen, Potassium, and Phosphorous are obtained from the soil and are the primary macronutrients[1]. Calcium, Magnesium, and Sulfur are the secondary macronutrients needed in lesser quantity. The micronutrients[2], needed in very small quantities and toxic in large quantities, include Iron, Manganese, Copper, Zinc, Boron, and Chlorine. A complete fertilizer provides all three primary macronutrients and some of the secondary and micronutrients. The label of the fertilizer will list numbers, for example 5-10-5, which refers to the percentage by weight of the primary macronutrients. Soils play a role 　　Soil is weathered, decomposed rock and mineral (geological) fragments mixed with air and water. Fertile soil contains the nutrients in a readily available form that plants require for growth. The roots of the plant act as miners moving through the soil and bringing needed minerals into the plant roots. 　　Plants use these minerals in: 　　1. Structural components in carbohydrates and proteins 　　2. Organic molecules used in metabolism[3], such as the Magnesium in chlorophyll and the Phosphorous found in ATP. 　　3. Enzyme activators like potassium, which activates possibly fifty enzymes. 　　4. Maintaining osmotic[4] balance. Mycorrhizae[5], bacteria, and minerals 　　Plants need nitrogen for many important biological molecules including nucleotides and proteins. However, the ni-	[1]大量养料,大量营养素 [2]微量养料,微量营养素 [3]新陈代谢 [4]渗透的,渗透性的 [5]菌根

续表

Text	Notes
trogen in the atmosphere is not in a form that plants can utilize. Many plants have a symbiotic[6] relationship with bacteria growing in their roots: organic nitrogen as rent for space to live. These plants tend to have root nodules[7] in which the nitrogen-fixing bacteria live. Development of a root nodule, a place in the roots of certain plants, most notably legumes[8] (the pea family), where bacteria live symbiotically with the plant. All the nitrogen in living systems was at one time processed by these bacteria, who took atmospheric nitrogen (N_2) and modified it to a form that living things could utilize (such as NO_3 or NO_4; or even as ammonia, NH_3 in the example shown below). Not all bacteria utilize the above route of nitrogen fixation. Many that live free in the soil, utilize other chemical pathways. Roots have extensions of the root epidemal cells known as root hairs[9]. While root hairs greatly enhance the surface area (hence absorption surface), the addition of symbiotic mycorrhizae fungi vastly increases the area of the root for absorbing water and minerals from the soil. Water and Mineral Uptake[10] Animals have a circulatory system[11] that transports fluids, chemicals, and nutrients around within the animal body. Some plants have an analogous system: the vascular system[12] in vascular plants; trumpet hyphae in bryophytes[13]. Root hairs are thin-walled extensions of the epidermal cells in roots. They provide increased surface area and thus more efficient absorption of water and minerals. Water and dissolved mineral nutrients enter the plant via two routes. Water and selected solutes only pass through the cell membrane of the epidermis of the root hair and then through plasmodesmata on every cell until they reach the xylem: in-	[6]共生的 [7]节结 [8]豆类,豆荚 [9]根毛 [10]摄取 [11]循环系统 [12]导管系统 [13]苔藓类植物

Text	Notes
tracellular route (apoplastic). Water and solutes enter the cell wall of the root hair and pass between the wall and plasma membrane until they encounter the endodermis, a layer of cells that they must pass through to enter the xylem: extracellular route. The endodermis has a strip of water-proof material (containing suberin[14]) known as the Casparian strip that forces water through the endodermal cell and in such a way regulates the amount of water getting to the xylem. Only when water concentrations inside the endodermal cell fall below that of the cortex parenchyma cells does water flow into the endodermis and on into the xylem. Xylem and Transport Xylem is the water transporting tissue in plants that is dead when it reaches functional maturity. Tracheids are long, tapered cells of xylem that have end plates on the cells that contain a great many crossbars. Tracheid walls are festooned with pits. Vessels, an improved form of tracheid, have no (or very few) obstructions (crossbars) on the top or bottom of the cell. The functional diameter[15] of vessels is greater than that of tracheids. Water is pulled up the xylem by the force of transpiration, water loss from leaves. Mature corn plants can each transpire four gallons of water per week. Transpiration rates in arid-region plants can be even higher. Water molecules are hydrogen bonded to each other. Water loss from the leaves causes diffusion of additional water molecules out of the leaf vein xylem, creating a tug on water molecules along the water columns within the xylem. This "tug" causes water molecules to rise up from the roots eventually to the leaves. The loss of water from the root xylem allows additional water to pass through the endodermis into the root xylem.	[14]软木脂 [15]直径

Text A 参考译文

植物与植物结构

一个植株有两种器官系统：①茎系统；②根系统。茎系统是指植株的地上部分，包括叶、芽、茎、花（如果这个植物有的话）和果实（如果这个植物有的话）等器官。根系统包括植株的地下部分，如根、块茎和根茎。

植物细胞是由分生组织形成的，然后发育形成多种类型的细胞，再形成各种组织。植物只有三种类型的组织：①真皮组织；②基本组织；③维管组织。真皮组织覆盖于草本植物的表面，是由表皮细胞和分泌蜡质角质层以防止水分丢失的填充细胞组成。基本组织是由大部分初级植物体组成，常见的有薄壁组织、厚角组织和厚壁组织细胞。维管组织在植株内运输养料、水分、激素和矿物质，包括木质部、韧皮部、薄壁组织和新生组织细胞。

各种植物细胞通过分生组织的有丝分裂产生。分生组织可以被认为是定位有丝分裂的场所。分生组织位于茎或根的顶端（顶端分生组织）或侧部，侧生分生组织随植株的长度呈圆柱体状扩展。新生组织是一种（通常）产生次级生长的侧生分生组织。次级生长既发生在木质组织，也发生在软木组织（但分别来自各自不同的次生分生组织）。

薄壁组织

薄壁细胞属于一种普通植物细胞类型，存在于成熟期。作为基本组织和维管组织的大部分，薄壁组织的作用是贮藏作用和光合作用。栅栏薄壁组织细胞是位于许多叶子表皮细胞下层的伸长细胞。海绵叶肉细胞在栅栏细胞下面1～2层。射线薄壁组织细胞位于木射线层，这些结构侧向运输物质进入木质茎。薄壁组织细胞也出现在维管束的木质部和韧皮部中。最大的薄壁组织细胞位于木髓区域，通常在玉米（玉蜀黍属）茎中，比维管束更大。在切片中常被染成绿色。

厚角组织

厚角组织细胞在植株中起支持作用。厚角组织细胞的特点是它的细胞壁增厚，成熟时厚角组织仍然是生活细胞。它们常常是维管束的一部分或位于有角茎的角隅处。厚角组织在许多切片中被染成红色。

厚壁组织

厚壁组织细胞也起着支持植株的作用。它们常常作为纤维束。厚壁组织细胞的特点是次生壁增厚。它们在成熟期死亡。它们与厚角组织一样在切片中常被染成红色。厚壁组织细胞常见的一种类型是纤维。

木质部

木质部是指植株一些特定细胞的木质（充满木质素）壁。木质部细胞主要把水分和无机盐从根部运送到叶部。如果说薄壁组织细胞确实属于"木质部"细胞的话。那么常被番红精-O染成红色的管胞和导管分子更可以确定是木质部细胞。管胞是两种细胞中更为原始的细胞,出现在早期维管植物中。管胞是长菱形细胞,由于其端部重叠成角可以形成细胞与细胞的连接。导管分子较短和较宽,缺乏端板。它们只出现在高等植物被子植物中。被子植物演化距今历史最短,但它演变成了一个植物大类群。

管胞比大部分导管长而窄,首先出现在化石记录中。导管出现较晚。管胞细胞的两端被条状物横切成斜角。导管的进化趋势是细胞变短,在细胞端壁上没有条状物。

韧皮部

韧皮部细胞把食物从叶子输送到植物其他部位。它们存在于成熟组织,常被染成绿色(用常绿色染色)。韧皮部细胞通常位于木质部的外表。韧皮部最常见的两种细胞是伴胞和筛胞。伴胞保留它们的细胞核,并控制邻近的筛胞。筛胞运输可溶性养料,如蔗糖。

表皮细胞

1. 表皮

表皮组织具有防止水分丢失的功能,并作为阻挡真菌类和其他侵入者的屏障。因此,表皮细胞之间连接非常紧密,几乎没有细胞间隙。为进一步减少水分丢失,许多植物有一蜡质表皮层沉积在表皮细胞上。

2. 保卫细胞

为促进气体在叶子、茎和果实内部之间的交换,植物有一系列被称为气孔的开口。显然这些开口允许气体交换,但是这是以水分丢失为代价的。保卫细胞是覆盖有气孔开口的豆形细胞。它们通过气孔调节水蒸气、氧气和二氧化碳的交换。

Text B 参考译文

花

花是呈涡旋形紧密排列的有很短节间的生殖组织和不育组织的总称。花的不育部分是萼片和花瓣。如果不育部分大小和形状上相似,就统称为被片。花的生殖部分是雄性花蕊(雄性的,统称为雄蕊)和心皮(心皮常被认为是雌蕊,雌性部分统称为雌蕊群)。

雄蕊群

雄蕊群的独立单位叫雄蕊,雄蕊由支撑花药的花丝组成。花药由4个小孢子囊组成,在小孢子囊中通过减数分裂产生小孢子(花粉)。

雄蕊被认为可代表变态孢子叶(在孢子囊上表面的孢子囊叶)。19世纪50年代James E. Canright的试验认为从有叶雄蕊的原始被子植物到其他"普通"雄蕊(百合)是进化级进。

花粉

花粉粒中含有植物雄配子(小孢子配子)体。花粉粒是位于花粉囊(小孢子囊)内缘

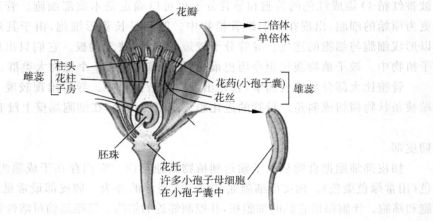

花的结构

的小孢子母细胞经过减数分裂产生的。花粉的外部是花粉粒外壁,它是由复合多糖和孢子花粉素组成的。在花粉内有 2 个(或至多 3 个)构成雄性配子的细胞。管细胞(也叫做管核)发育成为花粉管。1 个生殖细胞通过有丝分裂产生 2 个精细胞。生殖细胞的分裂可能在授粉前或授粉后进行。

雌蕊群

雌蕊由柱头、花柱和子房组成,子房内含一个或多个胚珠。这三种结构通常被合称为雌蕊或心皮。许多植物的雌蕊全部或部分融合了这些结构。

像雄蕊一样,心皮被认为是变态叶。I. W. Bailey 及其学生的研究指出了雌蕊群的进化次序是从原始被子植物心皮(像林仙属植物)到像百合属植物的"普通"心皮。

柱头与花柱

柱头具有接受器的功能:花粉在柱头上停留并在花粉管中发育。玉米穗丝是部分柱头和部分花柱。花柱将柱头与子房分隔开,分隔的距离因物种而异。

子房

子房包含一个或多个胚珠,胚珠包含一个雌配子,在被子植物也叫做胚囊。有些植物,如樱桃,只有一个子房,能产生两个胚珠。其中只有一个胚珠将发育成为种子。

配子

雄配子在花粉粒内发育。雌配子在子房内发育。开花植物配子体阶段形成的几个支持细胞在孢子体阶段可作为它们的营养。这正好与非维管植物地钱、苔藓和金鱼草的模式相反。

被子植物雄性配子有两个单倍体核(精核和管核),包含在花粉粒(或小孢子)外壁内。

有花植物的雌性配子在胚珠(大孢子)内发育,胚珠包含在花的雌蕊基部的子房内。在雌配子体中通常有8个(单倍体)细胞:a)1个卵细胞和2个在卵细胞两侧的助细胞(位于胚囊的珠孔端);b)2个位于胚囊中央的极核,以及3个反足细胞(在胚囊中卵细胞的另一端)。

双受精

花粉管通过柱头和花柱朝着子房内胚珠的方向生长,从而完成授粉过程。在花粉粒中的生殖细胞分裂,并产生两个精细胞沿花粉管向下移动。一旦花粉管头到达胚囊的珠孔端,花粉管通过卵细胞两侧的一个助细胞进入胚囊。一个精细胞与卵细胞融合,产生受精卵,以后将发育成为下一代孢子体。第二个精细胞与位于胚囊中心的两个极体融合,产生营养性的三倍体胚乳组织,为胚胎的生长和发育提供能量。

果实

子房壁在受精后发育成为果实。果实可以是肉质的、硬的,数量可以是多个或单个。

营养繁殖

许多植物进行无性繁殖。有些物种,如许多兰花,营养繁殖比种子繁殖更常见。块茎是地下肉质茎,如爱尔兰马铃薯。小叶是叶片的组成部分,它们可以发育成根,然后从植株中脱落,有效地进行无性繁殖。长匐茎是沿地表面生长的茎,它可以萌生出小植株,这种小植株植入地面发育成为新的独立的植株。

植物营养

与动物不同(它们通过食物获取营养),植物通过土壤和大气获取营养。植物以阳光作为能量来源,通过修饰光合作用形成的糖,能够制造自身需要的所有有机大分子。然而,植物还必须通过根系吸收各种矿物质供机体使用。

Unit 6

Text A

Organization of the Animal Body

Organs in animals are composed of a number of different tissue types.

Epithelial Tissue

Epithelial tissue covers body surfaces and lines body cavities. Functions include lining, protecting, and forming glands. Three types of epithelium occur:
- Squamous epithelium is flattened cells.
- Cuboidal epithelium is cube-shaped cells.
- Columnar epithelium consists of elongated cells.

Any epithelium can be simple or stratified. Simple epithelium has only a single cell layer. Stratified epithelium has more than one layer of cells. Pseudostratified epithelium is a single layer of cells so shaped that they appear at first glance to form two layers.

Functions of epithelial cells include:
- Movement of materials in, out, or around the body.
- Protection of the internal environment against the external environment.
- Secretion of a product.

Glands can be single epithelial cells, such as the goblet cells that line the intestine. Multicellular glands include the endocrine glands. Many animals have their skin composed of epithelium. Vertebrates have keratin in their skin cells to reduce water loss. Many other animals secrete mucus or other materials from their skin, such as earthworms do.

Connective Tissue

Connective tissue serves many purposes in the body:
- binding.
- supporting.

- protecting.
- forming blood.
- storing fats.
- filling space.

Connective cells are separated from one another by a non-cellular matrix. The matrix may be solid (as in bone), soft (as in loose connective tissue), or liquid (as in blood). Two types of connective tissue are Loose Connective Tissue (LCT) and Fibrous Connective Tissue (FCT). Fibroblasts are separated by a collagen fiber-containing matrix. Collagen fibers provide elasticity and flexibility. LCT occurs beneath epithelium in skin and many internal organs, such as lungs, arteries and the urinary bladder. This tissue type also forms a protective layer over muscle, nerves, and blood vessels.

Adipose tissue has enlarged fibroblasts storing fats and reduced intracellular matrix. Adipose tissue facilitates energy storage and insulation.

Fibrous Connective Tissue has many fibers of collagen closely packed together. FCT occurs in tendons, which connect muscle to bone. Ligaments are also composed of FCT and connect bone to bone at a joint.

Cartilage and bone are "rigid" connective tissues. Cartilage has structural proteins deposited in the matrix between cells. Cartilage is the softer of the two. Cartilage forms the embryonic skeleton of vertebrates and the adult skeleton of sharks and rays. It also occurs in the human body in the ears, tip of the nose, and at joints.

Bone has calcium salts in the matrix, giving it greater strength. Bone also serves as a reservoir (or sink) for calcium. Protein fibers provide flexibility while minerals provide elasticity. Two types of bone occur. Dense bone has osteocytes (bone cells) located in lacunae connected by canaliculi. Lacunae are commonly referred to as Haversian canals. Spongy bone occurs at the ends of bones and has bony bars and plates separated by irregular spaces. The solid portions of spongy bone pick up stress.

Blood is a connective tissue of cells separated by a liquid (plasma) matrix. Two types of cells occur. Red blood cells (erythrocytes) carry oxygen. White blood cells (leukocytes) function in the immune system. Plasma transports dissolved glucose, wastes, carbon dioxide and hormones, as well as regulating the water balance for the blood cells. Platelets are cell fragments that function in blood clotting.

Muscle Tissue

Muscle tissue facilitates movement of the animal by contraction of individual muscle cells (referred to as muscle fibers). Three types of muscle fibers occur in animals (the only taxonomic kingdom to have muscle cells):

- skeletal (striated).

- smooth.
- cardiac.

Muscle fibers are multinucleated, with the nuclei located just under the plasma membrane. Most of the cell is occupied by striated, thread-like myofibrils. Within each myofibril there are dense Z lines. A sarcomere (or muscle functional unit) extends from Z line to Z line. Each sarcomere has thick and thin filaments. The thick filaments are made of myosin and occupy the center of each sarcomere. Thin filaments are made of actin and anchor to the Z line.

Skeletal (striated) muscle fibers have alternating bands perpendicular to the long axis of the cell. These cells function in conjunction with the skeletal system for voluntary muscle movements. The bands are areas of actin and myosin deposition in the cells.

Smooth muscle fibers lack the banding, although actin and myosin still occur. These cells function in involuntary movements and/or autonomic responses (such as breathing, secretion, ejaculation, birth, and certain reflexes). Smooth muscle fibers are spindle shaped cells that form masses. These fibers are components of structures in the digestive system, reproductive tract, and blood vessels.

Cardiac muscle fibers are a type of striated muscle found only in the heart. The cell has a bifurcated (or forked) shape, usually with the nucleus near the center of the cell. The cells are usually connected to each other by intercalated disks.

Nervous Tissue

Nervous tissue functions in the integration of stimulus and control of response to that stimulus. Nerve cells are called neurons. Each neuron has a cell body, an axon, and many dendrites. Nervous tissue is composed of two main cell types: neurons and glial cells. Neurons transmit nerve messages. Glial cells are in direct contact with neurons and often surround them.

The neuron is the functional unit of the nervous system. Humans have about 100 billion neurons in their brain alone! While variable in size and shape, all neurons have three parts. Dendrites receive information from another cell and transmit the message to the cell body. The cell body contains the nucleus, mitochondria and other organelles typical of eukaryotic cells. The axon conducts messages away from the cell body.

New Words

cavity　　　　　　　　['kæviti]　　　　　　　　n. 腔;窝
epithelium　　　　　　[ˌepi'θi:ljəm]　　　　　　n. 上皮,上皮细胞

stratified	['strætifaid]	adj.	成层了的,分层的
intestine	[in'testin]	n.	肠
keratin	['kerətin]	n.	角蛋白
secrete	[si'kri:t]	vt.	分泌
mucus	['mju:kəs]	n.	黏液,胶
earthworm	['ə:θwə:m]	n.	蚯蚓
matrix	['meitriks]	n.	基质
fibroblast	['faibrəubla:st]	n.	纤维原细胞,成纤维细胞
collagen	['kɔlə,dʒən]	n.	胶原,成胶质
elasticity	[ilæs'tisiti]	n.	弹性,弹力
flexibility	[,fleksə'biliti]	n.	柔性,灵活性
artery	['ɑ:təri]	n.	动脉
intracellular	[,intrə'seljulə]	adj.	细胞内的
facilitate	[fə'siliteit]	vt.	帮助,使容易,促进
insulation	[,insju'leiʃən]	n.	绝缘
tendon	['tendən]	n.	腱,筋
ligament	['ligəmənt]	n.	韧带
cartilage	['kɑ:tilidʒ]	n.	软骨
bone	[bəun]	n.	骨骼
osteocyte	['ɔstiəsait]	n.	骨细胞
lacunae	[lə'kju:ni:]	n.	腔隙(lacuna 的复数)
canaliculi	[,kænə'likjuli:]	n.	小管,细管,微管(canaliculus 的复数)
erythrocyte	[i'riθrəusait]	n.	红细胞,红血球
leukocyte	['lju:kəsait]	n.	白细胞,白血球
platelet	['pleitlit]	n.	血小板
multinucleate	[,mʌlti'nju:kliit]	adj.	多核的
myofibril	[,maiəu'faibril]	n.	肌原纤维
sarcomere	['sɑ:kəmiə]	n.	肌(原纤维)节,肌小节
myosin	['maiəsin]	n.	肌浆球蛋白,阻凝蛋白
actin	['æktin]	n.	肌动蛋白,肌纤蛋白
cardiac	['kɑ:diæk]	adj.	心脏的
ejaculation	[i,dʒækju'leiʃən]	n.	射精
neuron	['njuərɔn]	n.	神经细胞,神经元
axon	['æksɔn]	n.	轴突
dendrite	['dendrait]	n.	枝状突起;树突

Phrases

epithelial tissue	上皮组织
squamous epithelium	扁平上皮
cuboidal epithelium	立方形上皮
columnar epithelium	柱形上皮
at first glance	乍一看,看一眼
goblet cells	杯状细胞
endocrine gland	内分泌腺
connective tissue	结缔组织
urinary bladder	膀胱
adipose tissue	脂肪组织
calcium salt	钙盐
spongy bone	海绵状骨骼
blood clotting	血液凝固
muscle tissue	肌肉组织
digestive system	消化系统
reproductive tract	生殖管
cardiac muscle fiber	心肌纤维
intercalated disks	闰盘
nervous tissue	神经组织
glial cell	神经胶质细胞
cell body	细胞体

Abbreviations

LCT (Loose Connective Tissue)　　疏松结缔组织
FCT (Fibrous Connective Tissue)　　纤维结缔组织

Notes

[1] Adipose tissue has enlarged fibroblasts storing fats and reduced intracellular matrix.

本句中,storing fats 是一个现在分词短语,作定语,修饰和限定 enlarged fibroblasts。该短语可以扩展为一个定语从句：which store fats。

[2] FCT occurs in tendons, which connect muscle to bone.

本句中,which connect muscle to bone 是一个非限定性定语从句,对 tendons 作进一步补充说明。

[3] Smooth muscle fibers lack the banding, although actin and myosin still occur.

本句中,although actin and myosin still occur 是一个让步状语从句,修饰主句的谓语 lack。

本句中,although 引导的让步状语从句可以放在句首,也可以放在句尾。主句中不能再用连接词 but,但可用副词 yet, nevertheless 等。请看下例:

Although it was so cold, he went out without an overcoat.
天气虽然很冷,他没有穿大衣就出去了。
Although they are poor, they are happy.
他们虽然很穷,但很快乐。

Exercises

【EX. 1】 根据课文内容,回答以下问题

1) How many types of tissues are discussed in the text? What are they?

2) What are the functions of epithelial tissue?

3) How many types of epithelium are there? What are they?

4) What are the functions of epithelial cells?

5) What purposes does connective tissue serve?

6) What are the two types of connective tissue?

7) What does cartilage form?

8) What does muscle tissue do?

9) How many types of muscle fibers occur in animals? What are they?

10) What does nervous tissue function?

【EX. 2】 根据下面的英文解释,写出相应的英文词汇

英　文　解　释	词　汇
Membranous tissue composed of one or more layers of cells separated by very little intercellular substance and forming the covering of most internal and external surfaces of the body and its organs.	
The portion of the alimentary canal extending from the stomach to the anus and in human beings and other mammals, consisting of two segments, the small intestine and the large intestine.	
A cell that gives rise to connective tissue.	
Generate and separate (a substance) from cells or bodily fluids	
A tough, elastic, fibrous connective tissue found in various parts of the body, such as the joints, outer ear, and larynx.	
The usually long process of a nerve fiber that generally conducts impulses away from the body of the nerve cell.	
A branched protoplasmic extension of a nerve cell that conducts impulses from adjacent cells inward toward the cell body.	
A minute, disklike cytoplasmic body found in the blood plasma of mammals that promotes blood clotting.	
Any of the impulse-conducting cells that constitute the brain, spinal column, and nerves, consisting of a nucleated cell body with one or more dendrites and a single axon.	

【EX. 3】 把下列句子翻译为中文

1) The bodies of humans and other mammals contain a cavity divided by the diaphragm into thoracic and abdominal cavities. The body's cells are organized into tissues, which are, in turn, organized into organs and organ systems.

2) Epithelial tissues include membranes that cover all body surfaces and glands.

3) Connective tissues are characterized by abundant extracellular materials in the matrix between cells. Connective tissue proper may be either loose or dense.

4) Skeletal muscles enable the vertebrate body to move. Cardiac muscle powers the heartbeat, while smooth muscles provide a variety of visceral functions.

5) There are different types of neurons, but all are specialized to receive, produce, and conduct electrical signals.

6) The vertebrate body is organized into cells, tissues, organs, and organ systems, which are specialized for different functions.

7) The four primary tissues of the vertebrate adult body—epithelial, connective, muscle, and nerve—are derived from three embryonic germ layers.

8) Smooth muscles are composed of spindle-shaped cells and are found in the organs of the internal environment and in the walls of blood vessels.

9) Neuroglia are supporting cells with various functions including insulating axons to accelerate an electrical impulse.

10) Skeletal and cardiac muscles are striated; skeletal muscles, however, are under voluntary control whereas cardiac muscle is involuntary.

【EX. 4】 把下列短文翻译为中文

Most animals have a body plan best described as a "tube-within-a-tube". This plan calls for two openings: one for food to enter the body (mouth), one for wastes to leave the body (anus). Animals with the "tube-within-a-tube" plan are 10% more efficient at digesting and absorbing their food than animals with the sac-like body plan. The tube-within-a-tube plan allows specialization of parts along the tube. The sac-like body plan has only one opening for both food intake and waste removal. Sac-like body plan animals do not have tissue specialization or development of organs.

Many, but not all, animals have three tissue layers as they develop embryologically: the endoderm, mesoderm, and ectoderm. Some animals, most notably sponges, lack these tissue layers. Cnidarians (coral and jellyfish) have only two of these layers. Flatworms, ribbon worms, etc. all have three tissue layers. Humans are triploblastic.

Text B

Some General Features of Animals

Animals are the eaters or consumers of the earth. They are heterotrophs and depend directly or indirectly on plants, photosynthetic protists (algae), or autotrophic bacteria for nourishment. Animals are able to move from place to place in search of food. In most, ingestion of food is followed by digestion in an internal cavity.

Multicellular Heterotrophs

All animals are multicellular heterotrophs. The unicellular heterotrophic organisms called Protozoa, which were at one time regarded as simple animals, are now considered to be members of the kingdom Protista.

Diverse in Form

Almost all animals (99%) are invertebrates, lacking a backbone. Of the estimated 10 million living animal species, only 42,500 have a backbone and are referred to as vertebrates. Animals are very diverse in form, ranging in size from ones too small to see with the naked eye to enormous whales and giant squids. The animal kingdom includes about 35 phyla, most of which occur in the sea. Far fewer phyla occur in fresh water and fewer still occur on land. Members of three phyla, Arthropoda (spiders and insects), Mollusca (snails), and Chordata (vertebrates), dominate animal life on land.

No Cell Walls

Animal cells are distinct among multicellular organisms because they lack rigid cell walls and are usually quite flexible. The cells of all animals but sponges are organized into structural and functional units called tissues, collections of cells that have joined together and are specialized to perform a specific function; muscles and nerves are tissues types, for example.

Active Movement

The ability of animals to move more rapidly and in more complex ways than members of other kingdoms is perhaps their most striking characteristic and one that is directly related to the flexibility of their cells and the evolution of nerve and muscle tissues. A remarkable form of movement unique to animals is flying, an ability that is well developed among both insects and vertebrates. Among vertebrates, birds, bats, and pterosaurs (now-extinct flying reptiles) were or are all strong fliers. The only terrestrial vertebrate group never to have had flying representatives is amphibians.

Sexual Reproduction

Most animals reproduce sexually. Animal eggs, which are nonmotile, are much larger than the small, usually flagellated sperm. In animals, cells formed in meiosis function directly as gametes. The haploid cells do not divide by mitosis first, as they do in plants and fungi, but rather fuse directly with each other to form the zygote.

Embryonic Development

Most animals have a similar pattern of embryonic development. The zygote first undergoes a series of mitotic divisions, called cleavage, and becomes a solid ball of cells, the morula, then a hollow ball of cells, the blastula. In most animals, the blastula folds inward at one point to form a hollow sac with an opening at one end called the blastopore. An embryo at this stage is called a gastrula. The subsequent growth and movement of the cells of the gastrula produce the digestive system, also called the gut or intestine. The details of embryonic development differ widely from one phylum of animals to another and often provide important clues to the evolutionary relationships among them.

The Classification of Animals

Two subkingdoms are generally recognized within the kingdom Animalia: Parazoa—animals that for the most part lack a definite symmetry and possess neither tissues nor organs, mostly comprised of the sponges, phylum Porifera; and Eumetazoa—animals that have a definite shape and symmetry and, in most cases, tissues organized into organs and organ systems. Although very different in structure, both types evolved from a common ancestral form and possess the most fundamental animal traits. All eumetazoas form distinct embryonic layers during development that differentiate into the tissues of the adult animal. Eumetazoas of the subgroup Radiata (having radial symmetry) have two layers, an outer ectoderm and an inner endoderm, and thus are called diploblastic. All other eumetazoas, the Bilateria (having bilateral symmetry), are triploblastic and produce a third layer, the mesoderm, between the ectoderm and endoderm. No such layers are present in sponges.

New Words

heterotroph	['hetərəutrɔf]	n. 异养生物
autotrophic	[ˌɔːtəu'trɔfik]	adj. 自养的, 无机营养的
ingestion	[in'dʒestʃən]	n. 摄取
digestion	[di'dʒestʃən]	n. 消化
protozoa	[ˌprəutəu'zəuə]	n. 原生动物 (protozoon 的复数)
invertebrate	[in'vəːtibrit]	n. 无脊椎动物
		adj. 无脊椎的
squid	[skwid]	n. 鱿鱼
Arthropoda	['ɑːθrəpədə]	n. 节足动物门
Mollusca	[məlʌskə]	n. 软体动物类

Chordata	[ˈkɔːdeitə]	n.	脊索动物类
terrestrial	[tiˈrestriəl]	adj.	陆地的,陆生的
flagellated	[ˈflædʒəleitid]	adj.	生有鞭毛的
zygote	[ˈzaigəut]	n.	受精卵,接合子,接合体
cleavage	[ˈkliːvidʒ]	n.	卵裂;裂解
morula	[ˈmɔːrjulə]	n.	桑椹胚
blastula	[ˈblæstjulə]	n.	囊胚
blastopore	[ˈblæstəupɔː]	n.	胚孔
gastrula	[ˈgæstrulə]	n.	原肠胚,胚囊,肠胚
gut	[gʌt]	n.	肠子
Parazoa	[ˌpærəˈzəuə]	n.	拟生动物,侧生动物
Eumetazoa	[juːmetəˈzəuə]	n.	真后生动物
Radiata	[ˈreidieitə]	n.	(无脊椎)辐射动物
ectoderm	[ˈektəudəːm]	n.	外胚层
endoderm	[ˈendəudəːm]	n.	内胚层
diploblastic	[ˌdipləuˈblæstik]	adj.	双胚层的
triploblastic	[ˌtripləuˈblæstik]	adj.	三胚层的
mesoderm	[ˈmesədəːm]	n.	中胚层

Phrases

photosynthetic protist	光合原生生物
in search of	寻找
haploid cell	单倍体细胞
embryonic development	胚胎发育

Exercises

【EX. 5】 根据课文内容,回答以下问题

1) What do animals depend directly or indirectly on for nourishment?

2) What were the unicellular heterotrophic organisms regarded as at one time? What are they now considered?

3) How many percent of animals are invertebrates?

4) How many phyla does the animal kingdom include?

5) Why are animal cells distinct among multicellular organisms?

6) Why are animals move more rapidly and in more complex ways than members of other kingdoms?

7) Which is larger, an animal egg or the small flagellated sperm?

8) How many subkingdoms are generally recognized within the kingdom Animalia?

9) What are Parazoas?

10) What are Eumetazoas?

Reading Material

Text	Notes
Animal Organ Systems and Homeostasis Animal organs are usually composed of more than one cell type. Organs perform a certain function. Most organs have functions in only one organ system. Organ systems are composed of organs, and perform a major function for the organism. Homeostasis Homeostasis[1] is the maintenance of a stable internal environment. Homeostasis is a term coined in 1959 to describe the physical and chemical parameters that an organism must maintain to allow proper functioning of its component cells, tissues, organs, and organ systems. Single-celled organisms are surrounded by their external environment. Most multicellular organisms have most of their cells protected from the external environment, having them surrounded by an aqueous internal environment. This internal environment	[1]体内平衡

续表

Text	Notes
must be maintained in such a state as to allow maximum efficiency. The ultimate control of homeostasis is done by the nervous system. Often this control is in the form of negative feedback loops[2]. Heat control is a major function of homeostatic conditions that involves the integration of skin, muscular, nervous, and circulatory systems. Multicellular organisms have a series of organs and organ systems that function in homeostasis. Changes in the external environment can trigger changes in the internal environment as a response. The Internal Environment 　　There are two types of extracellular fluids[3] in animals: 　　• the extracellular fluid that surrounds and bathes cells. 　　• plasma, the liquid component of the blood. Internal components of homeostasis: 　　1. Concentration of oxygen and carbon dioxide. 　　2. pH of the internal environment. 　　3. Concentration of nutrients and waste products. 　　4. Concentration of salt and other electrolytes[4]. 　　5. Volume and pressure of extracellular fluid. Control Systems 　　Open systems are linear[5] and have no feedback, such as a light switch. Closed Systems has two components: a sensor and an effector[6], such as a thermostat (sensor) and furnace (effector). Most physiological systems in the body use feedback to maintain the body's internal environment. Extrinsic 　　Most homeostatic systems are extrinsic: they are controlled from outside the body. Endocrine and nervous systems are the major control systems in higher animals. 　　The nervous system depends on sensors in the skin or sensory organs to receive stimuli and transmit a message to the spinal cord or brain. Sensory input is processed and a	[2]反馈环 [3]细胞外液 [4]电解质,电解液 [5]线状的;细长的 [6]神经效应器;受动器

Text	Notes

signal is sent to an effector system, such as muscles or glands, that effects the response to the stimulus.
 	
    The endocrine system is the second type of extrinsic control, and involves a chemical component to the reflex. Sensors detect a change within the body and send a message to an endocrine effector (parathyroid), which makes PTH. PTH is released into the blood when blood calcium levels are low. PTH causes bone to release calcium into the bloodstream, raising the blood calcium levels and shutting down the production of PTH.
 	
    Some reflexes have a combination of nervous and endocrine response. The thyroid gland[7] secretes thyroxin (which controls the metabolic rate) into the bloodstream. Falling levels of thyroxin stimulate receptors in the brain to signal the hypothalamus to release a hormone that acts on the pituitary gland to release thyroid-stimulating hormone (TSH) into the blood. TSH acts on the thyroid, causing it to increase production of thyroxin.
 	
Intrinsic
    Local, or intrinsic[8], controls usually involve only one organ or tissue. When muscles use more oxygen, and also produce more carbon dioxide, intrinsic controls cause dilation of the blood vessels allowing more blood into those active areas of the muscles. Eventually the vessels will return to "normal".
 	
Feedback Systems in Homeostasis
    Negative feedback control mechanisms (used by most of the body's systems) are called negative because the information caused by the feedback causes a reverse of the response. TSH[9] is an example: blood levels of TSH serve as feedback for production of TSH.
 	
    Positive feedback control is used in some cases. Input increases or accelerates the response. During uterine con-
	[7]甲状腺
	
	[8]本身的;内在的
	
	[9]促甲状腺激素

Text	Notes
tractions[10], oxytocin is produced. Oxytocin causes an increase in frequency and strength of uterine contractions. This in turn causes further production of oxytocin[11], etc. Homeostasis depends on the action and interaction of a number of body systems to maintain a range of conditions within which the body can best operate. Body Systems and Homeostasis 　　Eleven major organ systems are present within animals, although some animals lack one or more of them. The vertebrate body has two cavities: the thoracic[12], which contains the heart and lungs; and the abdominal[13], which contains digestive organs. The head, or cephalic region, contains four of the five senses as well as a brain encased in the bony skull. These organ systems can be grouped according to their functions. 　　Muscular System allows movement and locomotion. The muscular system produces body movements, body heat, maintains posture, and supports the body. Muscle fibers are the main cell type. Action of this system is closely tied to that of the skeletal system. 　　Skeletal System provides support and protection, and attachment points for muscles. The skeletal system provides rigid framework for movement. It supports and protects the body and body parts, produces blood cells, and stores minerals. 　　Skin or Integument[14] is the outermost protective layer. It prevents water loss from and invasion of foreign microorganisms and viruses into the body. There are three layers of the skin. The epidermis is the outer, thinner layer of skin. Basal cells continually undergo mitosis. Skin is waterproof because keratin, a protein is produced. The next layer is the dermis a layer of fibrous connective tissue. Within the dermis many structures are located, such as sweat glands, hair follicles and oil glands. The subcutaneous layer is composed	[10]子宫收缩 [11]缩宫素 [12]胸的 [13]腹部的 [14]皮肤,外皮

Text	Notes
of loose connective tissue. Adipose tissue occurs here, serving primarily for insulation. Nerve cells run through this region, as do arteries and veins. Respiratory System moves oxygen from the external environment into the internal environment; also removes carbon dioxide. The respiratory system exchanges gas between lungs (gills in fish) and the outside environment. It also maintains pH of the blood and facilitates exchange of carbon dioxide and oxygen. Digestive System digests and absorbs food into nutrient molecules by chemical and mechanical breakdown; eliminates solid wastes into the environment. Digestion is accomplished by mechanical and chemical means, breaking food into particles small enough to pass into bloodstream. Absorbing of food molecules occurs in the small intestine and sends them into circulatory system. The digestive system also recycles water and reclaims vitamins from food in the large intestine. Circulatory System transports oxygen, carbon dioxide, nutrients, waste products, immune components, and hormones. Major organs include the heart, capillaries, arteries, and veins. The lymphatic system[15] also transports excess fluids to and from circulatory system and transports fat to the heart. Immune System defends the internal environment from invading microorganisms and viruses, as well as cancerous cell growth. The immune system provides cells that aid in protection of the body from disease via the antigen/antibody response. A variety of general responses are also part of this system. Excretory System[16] regulates volume of internal body fluids as well as eliminates metabolic wastes from the internal environment. The excretory system removes organic wastes from the blood, accumulating wastes as urea in the kidneys. These wastes are then removed as urine. This system is also responsible for maintaining fluid levels.	[15]淋巴系统 [16]排泄系统

续表

Text	Notes
Nervous System coordinates and controls actions of internal organs and body systems. Memory, learning, and conscious thought are a few aspects of the functions of the nervous system. Maintaining autonomic functions such as heartbeat, breathing, control of involuntary muscle actions are performed by some of the parts of this system. Endocrine System works with the nervous system to control the activity internal organs as well as coordinating long-range response to external stimuli. The endocrine system secretes hormones that regulate body metabolism, growth, and reproduction. These organs are not in contact with each other, although they communicate by chemical messages dumped into the circulatory system. Reproductive System is mostly controlled by the endocrine system, and is responsible for survival and perpetuation of the species. Elements of the reproductive system produce hormones (from endocrine control) that control and aid in sexual development. Organs of this system produce gametes that combine in the female system to produce the next generation (embryo).	

Text A 参考译文

动物体组织

动物的器官是由不同类型的组织组成的。

上皮组织

上皮组织覆盖身体表面并顺体腔排列,有支撑功能、保护功能和形成腺体的功能。有3种类型的上皮组织:
- 扁平上皮是扁平状的细胞。
- 立方上皮是立方体状细胞。
- 柱状上皮由伸长的细胞组成。

上皮组织分为单层上皮组织和复层上皮组织两大类。单层上皮仅有一层单一细胞。复层上皮有一层以上的细胞。假复层上皮有许多单一层细胞,但从形状上乍看像两层。

上皮细胞的功能包括:

- 物质的进出或在体内移动。
- 保护内环境,抵御外环境。
- 物质的分泌。

腺体可以是单层上皮细胞,如顺肠道排列的杯状腺。多细胞腺包括内分泌腺。许多动物的皮肤由上皮组织组成。脊椎动物皮肤内有角蛋白以减少水分流失。还有许多其他动物,如蚯蚓的皮肤分泌黏液或其他物质。

结缔组织

结缔组织在身体内有许多作用:
- 黏合。
- 支持。
- 保护。
- 形成血液。
- 储存脂肪。
- 填充空间。

非细胞基质将结缔组织细胞彼此分开。这种基质可以是坚硬的(如在骨内)、柔软的(如在疏松结缔组织内)或液体的(如在血液中)。结缔组织分为两类:疏松结缔组织(LCT)和致密结缔组织(FCT)。纤维细胞被一种纤维基质胶原质分开。胶原质纤维提供弹性和柔性。皮肤和许多内脏器官(如肺、动脉和膀胱)上皮组织之下有 LCT 组织。这种组织类型也在肌肉、神经和血管上形成保护层。

脂肪组织由变长的能贮存脂肪的纤维原细胞和少量的细胞内基质组成。脂肪组织具有能量贮藏与绝缘功能。

致密结缔组织内含大量的密集的胶原纤维。在连接肌肉与骨骼的腱中有 FCT。在关节处连接骨骼与骨骼的韧带也由 FCT 组成。

软骨与骨骼是"刚性的"结缔组织。软骨是沉积在细胞间的基质内的结构蛋白。它比骨骼软,形成脊椎动物的胚胎骨架及鲨鱼和海星的成年骨架。在人类的耳朵、鼻尖和关节处也有软骨。

骨骼基质中的钙盐增大了骨骼的强度。骨骼也作为身体的钙库(或钙池)。矿物质提供刚性,蛋白纤维提供弹性。骨骼有两种类型,致密骨骼是位于由微管连接的腔隙内的骨细胞。腔隙通常也叫做哈弗氏管。海绵状骨骼位于骨骼的端部,有被不规则空间分隔开的骨条和骨板。海绵状骨骼的固体部分承受压力。

血液是一群被一种液体(血浆)基质分开的结缔组织细胞。血液细胞有两种:红细胞(红血球)运送氧气。白细胞(白血球)具有免疫功能。血浆传送分解的葡萄糖、废物、二氧化碳和激素,同时调节血细胞的水平衡。血小板是具有血液凝固作用的细胞碎片。

肌肉组织

肌肉组织通过肌肉细胞(又叫肌肉纤维)的收缩促进动物的运动。动物体内有 3 种类型的肌肉纤维(仅根据肌肉细胞分类):

- 骨骼肌。
- 平滑肌。
- 心肌。

  肌肉纤维是多核的,多个核位于细胞膜之下。这种细胞的大部分被条状、线状的肌原纤维所占据。在每一个肌原纤维内有密集的 Z 带。肌原纤维节(或肌功能单位)从一个 Z 带延伸到另一个 Z 带。每一个肌原纤维节都有粗丝和细丝。粗丝带由肌球蛋白组成,位于肌原纤维节的中心。细丝由肌动蛋白组成,固定在 Z 带。

  骨骼肌(条纹肌)纤维有垂直于细胞长轴的交叉条带。这些细胞的功能与骨骼系统协同引起肌肉随意运动。这些条带是肌动蛋白和肌球蛋白在细胞中的沉积。

  平滑肌有肌动蛋白和肌球蛋白,但没有条带。这些细胞在非随意运动和(或)自律反应(如呼吸、分泌、射精、分娩及某些反射)中起作用。平滑肌纤维是纺锤状细胞的聚集。这些纤维是消化系统、生殖管和血管的组成成分。

  心肌纤维是一种仅仅在心脏中发现的横纹肌。这种细胞呈分叉形状,细胞核通常位于细胞中央。这些细胞通常由闰盘连接起来。

神经组织

  神经组织的功能是对刺激进行分析与综合,以及控制机体对这些刺激作出相应的反应。神经细胞也叫神经元。每一个神经元有一个叫轴突的细胞体和许多树突。神经组织主要由两类细胞组成:神经元和神经胶质细胞。神经元传送神经信息。神经胶质细胞与轴突直接接触,并常常包围轴突。

  神经元是神经系统的功能单位。人类仅在大脑内就有大约 1000 亿个神经元。虽然大小和形状各异,但所有的神经元都由三个部分组成。树突从另一个细胞接受信息并传送信息到细胞体。细胞体包含有细胞核、线粒体和真核细胞的其他细胞器。轴突传播细胞体的信息。

Text B 参考译文

动物的一般特征

  动物是地球生态系统中的消费者。它们是异养生物,直接或间接依赖植物、光合原生生物(藻类),或者自养细菌提供营养。动物能到处寻找食物。在多数情况下,食物的摄取后紧接着的是食物在消化道内的消化。

多细胞异养生物

  所有的动物都是多细胞异养生物。单细胞异养生物又叫原生动物,原生动物曾被认为是最简单的动物,现在被认为是原生生物界的成员。

形状的多样性

  几乎所有的动物(99%)都是没有脊柱的无脊椎动物。估计 1000 万种活着的动物中,仅仅只有 42 500 种有脊椎,被称为脊椎动物。动物在形状、大小上差别非常大,从小到用肉眼看不到的动物,到巨大的鲸鱼和庞大的鱿鱼。动物界分为 35 门,大部分生活在海洋中。淡水动物门比海水动物门少得多,陆地动物门更少。节足动物门(蜘蛛和昆虫)、软体动物门(蜗牛)和脊索动物门这三个门中的动物类群是陆地上主要的动物类群。

没有细胞壁

在多细胞机体中,动物细胞比较特殊,因为它们没有坚硬的细胞壁,且通常十分柔软。除海绵外,所有动物细胞都组成组织,组织是机体结构和功能的基本单位。组织是结合在一起并履行特殊功能的许多细胞的集合,例如肌肉和神经就是组织类型。

主动运动

动物有比其他界生物以更迅速、以更复杂的方式进行运动的能力,这也许是它们最显著的特征,这与它们细胞的柔软性以及神经与肌肉组织进化有直接关系。动物特有的一种值得注意的运动形式是飞翔,它是昆虫和脊椎动物发育良好的一项能力。在脊椎动物中,鸟、蝙蝠和翼龙(已灭绝的会飞的爬行动物)曾经是或现在是强健的飞翔者。从来不会飞翔的陆地脊椎代表动物只有两栖动物。

有性繁殖

多数动物为有性繁殖。不会运动的动物卵子比小的、带有鞭毛的精子大得多。动物减数分裂中形成的细胞直接具有配子功能。就像它们在植物和真菌中一样,单倍体细胞不会先在有丝分裂中分裂,而是直接相互融合成为受精卵。

胚胎发育

多数动物有一种类似的胚胎发育模式。受精卵首先经受一系列的有丝分裂,即卵裂,变成一个实心细胞球(桑椹胚),然后成为一个有空的细胞球(囊胚)。在多数动物中,囊胚在一点上折叠内陷形成一个一头开口的(叫做胚孔)的空囊。这个阶段的胚胎叫做原肠胚。接下来,原肠胚细胞的生长和运动产生了消化系统,也叫消化道或肠道。胚胎发育在不同的门之间有很大的不同,常常为它们之间进化关系提供线索。

动物的分类

通常认为动物界内有两个亚门:

大部分侧生动物身体不对称,无组织和器官,主要由海绵、多孔动物门组成;真后生动物有固定的形状,身体对称,多数情况下,组织形成器官和系统。虽然在结构上很不相同,两种类型都从一个共同的祖先进化而来,并且都具有大部分动物的基本特性。所有真后生动物在发育期间形成独特的胚胎层,分别分化为成年动物的不同组织。真后生动物的无脊椎辐射动物亚群(具有辐射性对称)有两个胚层,一个外部的外胚层和一个内部的内胚层,这被称为双胚层。所有其他真后生动物、双侧对称(具有左右对称)动物是三胚层动物,即在外胚层和内胚层之间有一个中胚层。而海绵动物则不形成这样的胚层。

Unit 7

Text A

Ecosystems

Ecosystems include both living and nonliving components. These living, or biotic, components include habitats and niches occupied by organisms. Nonliving, or abiotic, components include soil, water, light, inorganic nutrients, and weather. An organism's place of residence, where it can be found, is its habitat. A niche is often viewed as the role of that organism in the community, factors limiting its life, and how it acquires food.

Producers, a major niche in all ecosystems, are autotrophic, usually photosynthetic, organisms. In terrestrial ecosystems, producers are usually green plants. Freshwater and marine ecosystems frequently have algae as the dominant producers.

Consumers are heterotrophic organisms that eat food produced by another organism. Herbivores are a type of consumer that feeds directly on green plants (or another type of autotroph). Since herbivores take their food directly from the producer level, we refer to them as primary consumers. Carnivores feed on other animals (or another type of consumer) and are secondary or tertiary consumers. Omnivores, the feeding method used by humans, feed on both plants and animals. Decomposers are organisms, mostly bacteria and fungi that recycle nutrients from decaying organic material. Decomposers break down detritus, nonliving organic matter, into inorganic matter. Small soil organisms are critical in helping bacteria and fungi shred leaf litter and form rich soil.

Even if communities do differ in structure, they have some common uniting processes such as energy flow and matter cycling. Energy flows move through feeding relationships. The term ecological niche refers to how an organism functions in an ecosystem. Food webs, food chains, and food pyramids are three ways of representing energy flow.

Producers absorb solar energy and convert it to chemical bonds from inorganic nutrients taken from environment. Energy content of organic food passes up food chain; eventually all energy is lost as heat, therefore requiring continual input. Original inorganic elements are mostly returned to soil and producers; can be used again by producers and no new input is required.

The flow of energy through an ecosystem. Energy flow in ecosystems, as with all other energy, must follow the two laws of thermodynamics. Recall that the first law states that energy is neither created nor destroyed, but instead changes from one form to another (potential to kinetic). The second law mandates that when energy is transformed from one form to another, some usable energy is lost as heat. Thus, in any food chain, some energy must be lost as we move up the chain.

The ultimate source of energy for nearly all life is the Sun. Recently, scientists discover an exception to this once unchallenged truism. Communities of organisms around ocean vents begin their food chain with chemosynthetic bacteria that oxidize hydrogen sulfide generated by inorganic chemical reactions inside the Earth's crust. In this special case, the source of energy is the internal heat engine of the Earth.

Food chains indicate who eats whom in an ecosystem. Represent one path of energy flow through an ecosystem. Natural ecosystems have numerous interconnected food chains. Each level of producer and consumers is a trophic level. Some primary consumers feed on plants and make grazing food chains; others feed on detritus.

The population size in an undisturbed ecosystem is limited by the food supply, competition, predation, and parasitism. Food webs help determine consequences of perturbations: if titmice and vireos fed on beetles and earthworms, insecticides that killed beetles would increase competition between birds and probably increase predation of earthworms, etc.

The trophic structure of an ecosystem forms an ecological pyramid. The base of this pyramid represents the producer trophic level. At the apex is the highest level consumer, the top predator. Other pyramids can be recognized in an ecosystem. A pyramid of numbers is based on how many organisms occupy each trophic level. The pyramid of biomass is calculated by multiplying the average weight for organisms times the number of organisms at each trophic level. An energy pyramid illustrates the amounts of energy available at each successive trophic level. The energy pyramid always shows a decrease moving up trophic levels because:

- Only a certain amount of food is captured and eaten by organisms on the next trophic level.
- Some of the food that is eaten cannot be digested and exits digestive tract as undigested waste.
- Only a portion of digested food becomes part of the organism's body; rest is used as source of energy.
- Substantial portion of food energy goes to build up temporary ATP in mitochondria that is then used to synthesize proteins, lipids, carbohydrates, fuel contraction of muscles, nerve conduction, and other functions.
- Only about 10% of the energy available at a particular trophic level is incorporated

into tissues at the next level. Thus, a larger population can be sustained by eating grain than by eating grain-fed animals since 100 kg of grain would result in 10 human kg but if fed to cattle, the result, by the time that reaches the human is a paltry 1 human kg!

A food chain is a series of organisms each feeding on the one preceding it. There are two types of food chain: decomposer and grazer. Grazer food chains begin with algae and plants and end in a carnivore. Decomposer chains are composed of waste and decomposing organisms such as fungi and bacteria.

Food chains are simplifications of complex relationships. A food web is a more realistic and accurate depiction of energy flow. Food webs are networks of feeding interactions among species.

The food pyramid provides a detailed view of energy flow in an ecosystem. The first level consists of the producers (usually plants). All higher levels are consumers. The shorter the food chain the more energy is available to organisms.

Most humans occupy a top carnivore role, about 2% of all calories available from producers ever reach the tissues of top carnivores. Leakage of energy occurs between each feeding level. Most natural ecosystems therefore do not have more than five levels to their food pyramids. Large carnivores are rare because there is so little energy available to them atop the pyramid.

Food generation by producers varies greatly between ecosystems. Net primary productivity (NPP) is the rate at which producer biomass is formed. Tropical forests and swamps are the most productive terrestrial ecosystems. Reefs and estuaries are the most productive aquatic ecosystems. All of these productive areas are in danger from human activity. Humans redirect nearly 40% of the net primary productivity and directly or indirectly use nearly 40% of all the land food pyramid. This energy is not available to natural populations.

New Words

biotic	[bai'ɔtik]	adj.	生物的,生命的
habitat	['hæbitæt]	n.	生活环境,产地、栖息地
niche	[nitʃ]	n.	小生境;生态位
abiotic	[ˌeibai'ɔtik]	adj.	无生命的,非生物的
community	[kə'mju:niti]	n.	群落
marine	[mə'ri:n]	adj.	海的;海中的
herbivore	['hə:bivɔ:]	n.	食草动物
carnivore	['kɑ:nivɔ:]	n.	食肉动物

tertiary	[ˈtəːʃəri]	adj.	第三的；第三位的；第三级的
omnivore	[ɒmˈnivɔː]	n.	杂食动物
decomposer	[ˌdiːkəmˈpəuzə]	n.	分解者
detritus	[diˈtraitəs]	n.	残渣，碎屑；碎石
kinetic	[kiˈnetik]	adj.	动的；动力学的，运动的
potential	[pəˈtenʃəl]	adj.	势的，位的
truism	[ˈtruːizm]	n.	公认的真理
chemosynthetic	[ˈkeməusinˈθetik]	adj.	化能合成的
oxidize	[ˈɒksiˌdaiz]	v.	氧化
trophic	[ˈtrɒfik]	adj.	营养的，有关营养的
predation	[priˈdeiʃən]	n.	捕食
parasitism	[ˈpærəsaitizəm]	n.	寄生（状态）
perturbation	[ˌpəːtəːˈbeiʃən]	n.	干扰，混乱
titmice	[ˈtitmais]	n.	山雀（titmouse 的复数）
vireo	[ˈviriəu]	n.	燕雀
beetle	[ˈbiːtl]	n.	甲虫
insecticide	[inˈsektisaid]	n.	杀虫剂
apex	[ˈeipeks]	n.	顶点，最高点
biomass	[ˈbaiəumæs]	n.	生物量
paltry	[ˈpɔːltri]	adj.	无价值的；微不足道的
grazer	[ˈgreizə]	n.	食草者；食植物者
depiction	[diˈpikʃən]	n.	描写，叙述
swamp	[swɒmp]	n.	沼泽，湿地
reef	[riːf]	n.	暗礁，珊瑚礁
estuary	[ˈestjuəri]	n.	河口；江口；入海口

Phrases

heterotrophic organism	异养生物
feed on	以……为食，以……为能源
inorganic matter	无机物
energy flow	能量流动，能流
matter cycling	物质循环
feeding relationship	采食关系
ecological niche	生态位，生态龛位
food web	食物网
food chain	食物链

food pyramids	食物金字塔
solar energy	太阳能
convert … to …	把……转换成……
tropical forest	热带森林
hydrogen sulfide	硫化氢
aquatic ecosystem	水体生态系统

Abbreviations

NPP (Net Primary Productivity)　　净初级生产力

Notes

[1] Consumers are heterotrophic organisms that eat food produced by another organism.

　　本句中,that eat food produced by another organism 是一个定语从句,修饰和限定 heterotrophic organisms。在该从句中,produced by another organism 是一个过去分词短语,作定语,修饰 food。

[2] Omnivores, the feeding method used by humans, feed on both plants and animals.

　　本句中,the feeding method used by humans 是一个名词短语,做 Omnivores 的同位语,对其作进一步补充说明。在该短语中,used by humans 是一个过去分词短语,作定语,修饰 the feeding method。

[3] Communities of organisms around ocean vents begin their food chain with chemosynthetic bacteria that oxidize hydrogen sulfide generated by inorganic chemical reactions inside the Earth's crust.

　　本句中,that oxidize hydrogen sulfide generated by inorganic chemical reactions inside the Earth's crust 是一个定语从句,修饰和限定 chemosynthetic bacteria。在该从句中,generated by inorganic chemical reactions inside the Earth's crust 是一个过去分词短语,作定语,修饰 hydrogen sulfide。

[4] At the apex is the highest level consumer, the top predator.

　　本句是一个倒装句,正常语序为:The highest level consumer, the top predator, is at the apex. the top predator 是对 the highest level consumer 的补充说明。

Exercises

【EX. 1】　根据课文内容,回答以下问题

1) What do ecosystems include?

2) What do living and nonliving components include respectively?

3) What does the term ecological niche refers to?

4) What are the three ways of representing energy flow?

5) What does the first law state?

6) What does the second law mandate?

7) How is the pyramid of biomass calculated?

8) What is a food chain?

9) How many types of food chain are there? What are they?

10) What are food webs?

【EX. 2】 根据下面的英文解释,写出相应的英文词汇

英 文 解 释	词 汇
The particular area within a habitat occupied by an organism.	
An animal that feeds chiefly on plants.	
An animal that feeds chiefly on flesh, such as dogs, cats, bears.	
The area or type of environment in which an organism or ecological community normally lives or occurs.	
The total mass of living matter within a given unit of environmental area.	
A chemical substance used to kill insects.	
Of the synthesis of carbohydrate from carbon dioxide and water using energy obtained from the chemical oxidation of simple inorganic compounds.	
The characteristic behavior or mode of existence of a parasite or parasitic population.	

英　文　解　释	词　　汇
An animal that feeds on everything available.	
The part of the wide lower course of a river where its current is met by the tides.	

【EX. 3】 把下列句子翻译为中文

1) Nitrogen becomes available to organisms almost entirely through the metabolic activities of bacteria, some free-living and others which live symbiotically in the roots of legumes and other plants.

2) Phosphates are relatively insoluble and are present in most soils only in small amounts. They are often so scarce that their absence limits plant growth.

3) Energy passes through ecosystems, a good deal being lost at each step.

4) Primary productivity occurs as a result of photosynthesis, which is carried out by green plants, algae, and some bacteria. Secondary productivity is the production of new biomass by heterotrophs.

5) Considerable energy is lost at each stage in food chains, which limits their length. In general, more productive food chains can support longer food chains.

6) Because of the linked nature of food webs, species on different trophic levels will effect each other, and these effects can promulgate both up and down the food web.

7) Species richness promotes ecosystem productivity and is fostered by spatial heterogeneity and stable climate.

8) Organisms use a variety of physiological, morphological, and behavioral mechanisms to adjust to environmental variation. Over time, species evolve adaptations to living in different environments.

9) These communities, which occur in regions of similar climate, are much the same wherever they are found. Variation in annual mean temperature and precipitation are good predictors of what biome will occur there.

10) Circulation of ocean water redistributes heat, warming the western side of continents. Disturbances in ocean currents like El Niño can have profound influences on world climate.

【EX. 4】 把下列短文翻译为中文

Components and Boundaries of Ecosystem

Physical substances can include organic materials that were once alive, such as bits of wood from trees, rotting plant material, and animal wastes and dead organisms. The physical substance of an ecosystem also includes inorganic materials such as minerals, nitrogen, and water, as well as the overall landscape of mountains, plains, lakes, and rivers.

The organisms and the physical environment of an ecosystem interact with one another. The atmosphere, water, and soil allow life to flourish and limit what kind of life can survive. For example, a freshwater lake provides a home for certain fish and aquatic plants. Yet, the same lake would kill plants and animals adapted to a saltwater estuary.

Just as the environment affects organisms, organisms affect their environment. Lichens break down rock. Trees block sunlight, change the acidity and moisture content of soil, and release oxygen into the atmosphere. Elephants may uproot whole trees in order to eat their leaves, beavers dam streams and create meadows, and rabbits nibble grasses right down to the ground.

Ecosystems are not closed; in fact, an ecosystem's boundaries are usually fuzzy. A pond, for example, blends little by little into marsh, and then into a mixture of open meadow and brush. A stream brings nutrients and organisms from a nearby forest and carries away materials to other ecosystems. Even large ecosystems interact with other ecosystems. Seeds blow from place to place, animals migrate, and flowing water and air carry organisms—and their products and remains—from ecosystem to ecosystem.

All ecosystems taken together make up the biosphere, all living organisms on the earth and their physical environment. The biosphere differs from other ecosystems in having fixed boundaries. The biosphere covers the whole surface of the earth. It begins underground and extends into the highest reaches of the atmosphere.

Text B

Terrestrial Biomes

Tundra and Desert

The tundra and desert biomes occupy the most extreme environments, with little or no moisture and extremes of temperature acting as harsh selective agents on organisms that occupy these areas. These two biomes have the fewest numbers of species due to the stringent environmental conditions. In other words, not everyone can live there due to the specialized adaptations required by the environment.

Tropical Rain Forests

Tropical rain forests occur in regions near the equator. The climate is always warm (between 20℃ and 25℃) with plenty of rainfall (at least 190 cm/year). The rain forest is probably the richest biome, both in diversity and in total biomass. About 17 million hectares of rain forest are destroyed each year (an area equal in size to Washington state). Estimates indicate the forests will be destroyed (along with a great part of the Earth's diversity) within 100 years. Rainfall and climate patterns could change as a result.

Temperate Forests

The temperate forest biome occurs in south of the taiga in eastern North America, eastern Asia, and much of Europe. Rainfall is abundant (30-80 inches/year; 75-200 cm) and there is a well-defined growing season of between 140 and 300 days. The eastern United States and Canada are covered (or rather were once covered) by this biome's natural vegetation, the eastern deciduous forest. Dominant plants include beech, maple, oak; and other deciduous hardwood trees. Trees of a deciduous forest have broad leaves, which they lose in the fall and grow again in the spring.

Shrubland

The shrubland biome is dominated by shrubs with small but thick evergreen leaves that are often coated with a thick, waxy cuticle, and with thick underground stems that survive the dry summers and frequent fires. Shrublands occur in parts of South America, western Australia, central Chile, and around the Mediterranean Sea. Dense shrubland in California, where the summers are hot and very dry, is known as chaparral. This Mediterranean-type shrubland lacks an understory and ground litter, and is also highly flammable. The seeds of many species require the heat and scarring action of fire to induce germination.

Grasslands

Grasslands occur in temperate and tropical areas with reduced rainfall (10-30 inches per year) or prolonged dry seasons. Grasslands occur in the Americas, Africa, Asia, and Australia. Soils in this region are deep and rich and are excellent for agriculture.

Grasslands are almost entirely devoid of trees, and can support large herds of grazing animals. Natural grasslands once covered over 40 percent of the Earth's land surface. In temperate areas where rainfall is between 10 and 30 inches a year, grassland is the climax community because it is too wet for desert and too dry for forests.

Deserts

Deserts are characterized by dry conditions (usually less than 10 inches per year; 25 cm) and a wide temperature range. The dry air leads to wide daily temperature fluctuations from freezing at night to over 120 degrees during the day. Most deserts occur at latitudes of 30° N or 30° S where descending air masses are dry. Some deserts occur in the rainshadow of tall mountain ranges or in coastal areas near cold offshore currents. Plants in this biome have developed a series of adaptations (such as succulent stems, and small, spiny, or absent leaves) to conserve water and deal with these temperature extremes. Photosynthetic modifications (CAM) are another strategy to life in the drylands.

Taiga (Boreal Forest)

The taiga is a coniferous forest extending across most of the northern area of northern Eurasia and North America. This forest belt also occurs in a few other areas, where it has different names: the montane coniferous forest when near mountain tops; and the temperate rain forest along the Pacific Coast as far south as California. The taiga receives between 10 and 40 inches of rain per year and has a short growing season. Winters are cold and short, while summers tend to be cool. The taiga is noted for its great stands of spruce, fir, hemlock, and pine. These trees have thick protective leaves and bark. The needlelike (evergreen) leaves can withstand the weight of accumulated snow. Taiga forests have a limited understory of plants, and a forest floor covered by low-lying mosses and lichens. Conifers, alders, birch and willow are common plants; wolves, grizzly bears, moose, and caribou are common animals. Dominance of a few species is pronounced, but diversity is low when compared to temperate and tropical biomes.

Tundra

The tundra covers the northernmost regions of North America and Eurasia, about 20% of the Earth's land area. This biome receives about 20 cm (8-10 inches) of rainfall annually. Snow melt makes water plentiful during summer months. Winters are long and dark, followed by very short summers. Water is frozen most of the time, producing frozen soil, permafrost. Vegetation includes no trees, but rather patches of grass and shrubs; grazing musk ox, reindeer, and caribou exist along with wolves, lynx, and rodents. A few animals highly adapted to cold live in the tundra year-round (lemming, ptarmigan). During the summer the tundra hosts numerous insects and migratory animals. The ground is nearly completely covered with sedges and short grasses during the short summer. There are also plenty of patches of lichens and mosses. Dwarf woody shrubs flower and produce seeds quickly during the short growing season. The alpine tundra occurs above the timberline on mountain ranges, and may contain many of the same plants as the arctic tundra.

Aquatic Biomes

Conditions in water are generally less harsh than those on land. Aquatic organisms are buoyed by water support, and do not usually have to deal with desiccation. Despite covering 71% of the Earth's surface, areas of the open ocean are a vast aquatic desert containing few nutrients and very little life. Clearcut biome distinctions in water, like those on land, are difficult to make. Dissolved nutrients controls many local aquatic distributions. Aquatic communities are classified into: freshwater (inland) communities and marine (saltwater or oceanic) communities.

New Words

biome	['baiəum]	n.	生物群落,生物群系
tundra	['tʌndrə]	n.	冻原,苔原
moisture	['mɔistʃə]	n.	湿气;水分;水蒸气
stringent	['strindʒənt]	adj.	苛刻的
taiga	['teigə]	n.	针叶林带,针叶树林地带,泰加群落
abundant	[ə'bʌndənt]	adj.	充足的,大量的,丰富的
vegetation	[ˌvedʒi'teiʃən]	n.	植被,(总称)植物
beech	[bi:tʃ]	n.	山毛榉
maple	['meipl]	n.	槭树,枫树
oak	[əuk]	n.	栎树,橡树
deciduous	[di'sidʒuəs]	adj.	(每年)落叶的
hardwood	['hɑːdwud]	n.	阔叶树,硬木
shrubland	['ʃrʌblænd]	n.	灌木丛
shrub	[ʃrʌb]	n.	灌木,灌木丛
chaparral	[tʃæpə'ræl]	n.	丛林,茂密的树丛;沙巴拉群落
understory	['ʌndəˌstɔːri]	n.	林下叶层
flammable	['flæməbl]	adj.	易燃的,可燃性的

induce	[in'dju:s]	vt. 促使,导致
germination	[ˌdʒə:mi'neiʃən]	n. 萌芽,发生
fluctuation	[ˌflʌktju'eiʃən]	n. 波动,起伏
latitude	['lætitju:d]	n. 纬度,范围
montane	['mɔntein]	adj. 山区的
spruce	[spru:s]	n. 云杉
fir	[fə:]	n. 冷杉,枞树,杉木
hemlock	['hemlɔk]	n. 毒芹,铁杉
pine	[pain]	n. 松树
bark	[bɑ:k]	n. 树皮
conifer	['kəunifə]	n. 针叶树
alder	['ɔ:ldə]	n. 桤木
birch	[bə:tʃ]	n. 桦树,白桦
willow	['wiləu]	n. 柳树
moose	[mu:s]	n. 驼鹿
caribou	['kæribu:]	n. 北美驯鹿
reindeer	['reindiə]	n. 驯鹿
lynx	[liŋks]	n. 山猫,猞猁
rodent	['rəudənt]	n. 啮齿动物
lemming	['lemiŋ]	n. 旅鼠
sedge	[sedʒ]	n. 莎草
timberline	['timbəlain]	n. 树木线
aquatic	[ə'kwætik]	adj. 水生的
desiccation	[ˌdesi'keiʃən]	n. 干燥,失水
distribution	[distri'bju:ʃən]	n. 分布

Phrases

tropical rain forest	热带雨林
temperate forest	温带森林
deciduous forest	落叶林
the Mediterranean Sea	地中海(= the Mediterranean)
be devoid of	空的;缺乏的
succulent stem	肉质茎
Boreal Forest	北方森林
coniferous forest	针叶林
grizzly bear	灰熊

musk ox   麝鹿
be buoyed by   被……支撑

Exercises

【EX. 5】 根据课文内容，回答以下问题

1) What are the two biomes which have the fewest numbers of species due to the stringent environmental conditions?

2) Where do tropical rain forests occur? What is the climate there?

3) Where do the temperate forest biome occur? What about the rainfall there?

4) What is the shrubland biome is dominated by? Where do shrublands occur?

5) Where do grasslands occur? What about the soil in this region?

6) What are deserts characterized by?

7) What is the weather like in the taiga?

8) What is the taiga noted for?

9) What does the tundra cover? What is the rainfall this biome receives annually?

10) How many communities are aquatic communities classified into? What are they?

Reading Material

Text	Notes
Community and Ecosystem Dynamics A community is the set of all populations that inhabit a certain area. Communities can have different sizes and boundaries. These are often identified with some difficulty.	

Text	Notes
An ecosystem is a higher level of organization the community plus its physical environment. Ecosystems include both the biological and physical components affecting the community/ecosystem. We can study ecosystems from a structural[1] view of population distribution or from a functional[2] view of energy flow and other processes. Community Structure 　　Ecologists find that within a community many populations are not randomly distributed. The recognition that there is a pattern and process of spatial distribution of species is a major accomplishment of ecology. Two of the most important patterns are open community structure and the relative rarity of species within a community. 　　Do species within a community have similar geographic range and density peaks? If they do, the community is said to be a closed community, a discrete unit with sharp boundaries known as ecotones[3]. An open community, however, has its populations without ecotones and distributed more or less randomly. 　　In a forest, where we find an open community structure, there is a gradient[4] of soil moisture. Plants have different tolerances to this gradient land at different places along the continuum. Where the physical environment has abrupt[5] transitions, we find sharp boundaries developing between populations. For example, an ecotone develops at a beach separating water and land. 　　Open structure provides some protection for the community. Lacking boundaries, it is harder for a community to be destroyed in an all or nothing fashion. Species can come and go within communities over time, yet the community as a whole persists. In general, communities are less fragile and more flexible than some earlier concepts would suggest. 　　Most species in a community are far less abundant than the dominant species that provide a community its name: for example oak-hickory, pine, etc. Populations of just a few species are dominant within a community, no matter what community we examine. Resource partitioning[6] is thought to be the main cause for this distribution.	[1]构造的,结构的 [2]功能的 [3]群落交错区 [4]梯度 [5]突然的 [6]划分

Text	Notes
Classification of Communities 　　There are two basic categories of communities: terrestrial (land) and aquatic (water). These two basic types of community contain eight smaller units known as biomes. A biome is a large-scale category containing many communities of a similar nature, whose distribution is largely controlled by climate. 　　Terrestrial Biomes: tundra, grassland, desert, taiga, temperate forest, tropical forest. 　　Aquatic Biomes: marine, freshwater.	
Community Density and Stability[7] 　　Communities are made up of species adapted to the conditions of that community. Diversity[8] and stability help define a community and are important in environmental studies. Species diversity decreases as we move away from the tropics. Species diversity is a measure of the different types of organisms in a community (also referred to as species richness). Latitudinal diversity gradient refers to species richness decreasing steadily going away from the equator. A hectare[9] of tropical rain forest contains 40-100 tree species, while a hectare of temperate zone forest contains 10-30 tree species. In marked contrast, a hectare of taiga contains only a paltry 1-5 species. Habitat destruction in tropical countries will cause many more extinctions per hectare than it would in higher latitudes. 　　Environmental stability is greater in tropical areas, where a relatively stable/constant environment allows more different kinds of species to thrive[10]. Equatorial communities are older because they have been less disturbed by glaciers and other climate changes, allowing time for new species to evolve. Equatorial areas also have a longer growing season. 　　The depth diversity gradient is found in aquatic communities. Increasing species richness with increasing water depth. This gradient is established by environmental stability and the increasing availability of nutrients. 　　Community stability refers to the ability of communities to remain unchanged over time. During the 1950s and 1960s, stability	[7] 稳定性 [8] 差异，多样性 [9] 公顷 [10] 兴旺，繁荣，茁壮成长，旺盛

Text	Notes
was equated[11] to diversity: diverse communities were also stable communities. Mathematical modeling during the 1970s showed that increased diversity could actually increase interdependence among species and lead to a cascade effect[12] when a keystone species is removed. Thus, the relation is more complex than previously thought.	[11] 相等,等同 [12] 级联效应;瀑布效应

Text A 参考译文

生态系统

生态系统包括生物和非生物组分两部分。生命或生物的组分包括有机体的栖息地和小生境。非生物成分或非生命的组分包括土壤、水、光、无机营养素和气候。一种有机体居住的地方,或者说它能被发现的地方就是它的栖息地。而小生境通常是指有机体在生物群落中的地位、限制它生命的因素,以及它获得食物的方式。

生产者通常是指能进行光合作用的自养生物,它占据了所有生态系统中的主要生态位。在陆地生态系统中,生产者通常是绿色植物。在淡水和海水生态系统中,通常是藻类作为主要的生产者。

消费者是以其他生物为食物的异养生物。草食动物是直接以绿色植物(或另一种自养生物)为食物的一类消费者。由于草食动物直接以生产者为食物,我们称之为一级消费者。肉食动物以另一种动物(或另一种消费者)为食,称为二级消费者或三级消费者。杂食动物,即人类的采食方式,既吃植物又吃动物。分解者是指从腐败的有机物中获得营养素的多数的细菌和真菌。分解者将碎屑和非生命有机物分解为无机物质。土壤中的小生物在帮助细菌和真菌分解碎叶片形成肥沃的土壤中起决定性的作用。

尽管生物群落之间在结构上各不相同,它们在能量流动和物质循环方面有一些共同之处。生物通过采食关系进行能量流动。生态位反映了一种生物在一个生态系统中的作用。食物网、食物链和食物金字塔是描述能量流动的三种方式。

生产者吸收太阳能,并将从环境中得到的无机营养物质转化为化学能。有机食物中的能量通过食物链向上传递,最终都以热的形式损耗,所以需要持续提供能量。最初的无机元素大部分重新回到土壤和生产者中,能重新被生产者利用,不需要提供新能量。

能量通过生态系统进行流动。这种流动与所有其他能量一样,遵循热力学的两个基本定理。热力学第一定律告诉我们,能量既不能被创造也不能被消灭,但是可以从一种形式转变为另一种形式(势能到动能)。热力学第二定律指当能量从一种形式转变为另一种形式时,一些有用的能以热的形式被散发出去。这样,在任何食物链中,一些能量在食物链上传递过程中被消耗了。

几乎所有生命的最终能量来源都是太阳。最近,科学家发现这个曾经被认为是绝对真理的理论却有一个例外:海洋口生物群落的食物链开始于化能合成细菌,这种细菌能利用地壳内的无机化学反应产生氧化硫化氢。在这个特例中,能量的来源是地球内部的热能。

食物链显示了生态系统中谁吃谁的问题,描述了生态系统中能量流动的途径。自然生态系统中有无数的相互连接的食物链。每一个环节上的生产者与消费者都是一个营养级。有些初级消费者以植物为食,形成放牧食物链;其他初级消费者以腐屑为食,形成腐屑食物链。

在没有受干扰的生态系统中,种群的大小受食物供给、竞争、捕食和寄生状态的限制。食物网帮助确定干扰的结果:如果山雀和燕雀以甲虫和蚯蚓为食,那么杀死甲虫的杀虫剂将促进两种鸟类之间的竞争,同时可能会增加对蚯蚓的捕食等。

一个生态系统的营养结构形成一个生态金字塔。金字塔的最低层反映了生产者的营养水平。最顶端是最高级消费者——顶端的食肉动物。在一个生态系统中也可以出现其他金字塔。金字塔的数量是以占据了每一个营养级的生物的数量为基础的。金字塔的生物量是指在每一个营养级中的生物平均质量与生物数量的乘积。一个能量金字塔描述了每一个连续的营养级中可利用的能量值。这个能量金字塔显示:能量从一个营养级向上一个营养级的流动是逐渐减少的。因为:

- 只有一部分食物被下一个营养级的有机体所捕食。
- 一部分被摄入而没有消化的食物作为未消化的废物从消化道排出。
- 仅仅只有一部分已消化的食物转化为有机体的组成部分,其余的被作为能量的来源。
- 食物能量的一部分在线粒体中形成ATP,用于蛋白质、脂肪、碳水化合物的合成,肌肉的收缩,神经的传导和其他功能。
- 一个特定的营养级仅仅只有10%的可利用能被转变为下一个营养级的有机体组织。

因为100kg的谷物能转变成10kg的人体体重,而如果用100kg的谷物先喂牛,人再以牛为食,结果只能转变成不到1kg的人体体重,因此吃谷物的生物群体比吃以谷物为食的动物的生物群体要大。

一个食物链是一系列以上一级生物为食物的有机体组成的。有两种食物链:腐屑食物链和放牧食物链。放牧食物链以藻类和植物开始,以食肉动物结束。腐屑食物链由有机物的尸体和微生物(如真菌和细菌)组成。

食物链是复杂关系的简化。食物网是对能量流动更为真实和准确的描述。食物网是物种之间相互作用的采食网络。

食物金字塔是对一个生态系统能量流动的详细描述。第一级由生产者(通常是植物)组成。其余较高级都是消费者。食物链越短,给生物的有效能越多。

大多数人处于生态金字塔顶层的食肉动物层。来自生产者的可利用能大约有2%可形成顶层食肉动物的组织。每一级食物层之间都有能量流失。所以大多数自然生态系统中,食物链在食物金字塔中不会超过5层。大型食肉动物量很少,因为处于金字塔顶部的它们可利用能太少了。

生产者生产的食物量在不同生态系统中差异很大。净初级生产力(NPP)是指生产者形成生物量的比率。热带森林和沼泽是陆地生态系统的最大生产者。暗礁和河口是水体生态系统的最大生产者。所有这些生产区域正在受到人类活动的威胁。人类改变了约40%的净初级生产力,并直接或间接地利用陆地食物金字塔中约40%的能量。这些能量是自然种群的无效能。

Text B 参考译文

陆地生物群落

冻原与沙漠

  冻原和沙漠生物群落是最极端环境下的生物群落,它们处于几乎没有水以及极端温度的环境,那里的生物受到严格的自然选择。由于极端的环境条件,使这两个生物群落拥有最少的物种。换言之,由于那里的生物需要对这种极端环境具有特殊的适应性,并不是每一种生物都可以在那里生存。

热带雨林

  热带雨林生物群落位于赤道的附近。那里的气候总是温暖(气温在20℃到25℃之间)、雨量充足(年降雨量至少在190cm)。热带雨林可能是最丰富的生物群落,拥有的物种多样性最丰富,生物总量最多。每年约有1700万公顷的热带雨林遭受破坏(相当于华盛顿州大小)。估计森林(连同大部分的地球生物多样性)将在100年内被毁灭。结果,地球的降雨量和气候状况将会发生变化。

温带森林

  温带森林生物群落出现在北美洲东部的针叶林地带南部、亚洲东部和欧洲的大部分地区。在那里,雨量丰富(30～80英寸/年;75～200cm/年),良好生长季节在140～300天之间。美国东部和加拿大所覆盖(或者说曾经被覆盖)的这种生物群落的自然植被是东部落叶林。其主要优势植物包括山毛榉树、枫树、橡树,以及其他落叶阔叶树。落叶林树的阔叶秋季掉落,春季又重新生长。

灌木丛

  灌木丛生物群落主要是灌木,这些灌木的叶子为小而厚的常绿叶,绿叶表面通常覆有一层厚的蜡质外皮,灌木的茎为粗大的地下茎,使其能在干热的夏天和频繁的火灾下存活。灌木出现在南美洲的部分地区、澳大利亚西部、智利中部和地中海周围。加利福尼亚的夏天又热又干燥,密集的灌木被认为是丛林。这种地中海类型的灌木缺乏林下叶层和地面树叶,具高度易燃性。许多物种的种子需要火的加热和愈伤作用来诱导发芽。

草原

  草原出现在雨量较少(10～30英寸/年)或干燥季节较长的温带和热带地区,分布在美洲、非洲、亚洲和澳大利亚。这个区域的土壤深而肥沃,对农作物生长极其有利。草原上几乎没有树木,能容纳大量的放牧动物。自然草原曾经覆盖地球陆地表面的40%以上。在年雨量在10～30英寸的温带地区，因为那里的气候对于沙漠来说太湿，而对于雨林来说又太干，草原为顶极群落。

沙漠

沙漠的特点是气候干燥（通常雨量少于10英寸；25cm），温差范围大。干燥的空气导致日温度变化极大，从夜晚的严寒到白天的高于120华氏度以上的高温。大多数沙漠出现在北纬30°或南纬30°，在那里气团是干的。有些沙漠出现在高山山脉的少雨区或在靠近冷岸流的海岸区。这个群落的植物发展了保存水分以及适应极端温度的适应能力（如肉质茎和小叶、刺状叶或叶子稀少）。光合作用模式的变化（景天酸代谢光合作用）是适应干旱地生活的另一种策略。

针叶树林地带（北方森林）

针叶树林地带是分布于北欧亚大陆和北美洲地区的大部分北方地区的针叶树林。这些森林地带也出现在一些其他地区，但它有不同的名字：如在山顶附近的称为山区针叶树林；沿太平洋海岸，远至加利福尼亚南部的称为温带雨林。针叶树林地带年降雨量为10～40英寸，生长季节短。冬季冷而短，夏季凉爽。针叶树林以拥有大量的云杉、冷杉、铁杉和松树而著名。这些树种有厚的保护叶和树皮，以及针状（常绿）叶能承受积雪的重压。针叶林有一些林下叶层，同时森林地表覆盖着低矮的苔藓和地衣类植物。常见的植物是松树、桤桦树和柳树，常见的动物是狼、大灰熊、驼鹿和北美驯鹿。有一些优势物种，但较温带和热带生物群落的生物多样性少。

冻原

冻原出现北美洲和欧亚大陆的北极，约占地球陆地面积的20%。冻原生物群落每年降雨量约8～10英寸（20cm）。在夏季融雪产生大量的水。冬季长而黑暗，紧接着是非常短促的夏季。多数时间，水是冻结的，产生冻土、永久冻结带。植物中没有树，只点缀着一些草和灌木，麝牛、驯鹿和北美驯鹿与狼、猞猁和啮齿动物在那里生存。一些动物能高度适应长年冻土的寒冷生活（旅鼠、松鸡类）。在夏季，冻原地带生活着众多的昆虫和迁徙的动物。在短暂的夏季，土地几乎完全被莎草和矮草所覆盖；同时有大量的苔藓和地衣作点缀。在短暂的成长季节，低矮的木本灌木开花，并很快产生种子。高山冻原出现在高山林木线以上，其间分布的许多植物与北极冻原植物相同。

水生生物群落

水环境通常没有陆地环境那样恶劣。水生生物有水的支撑，不涉及干燥环境。尽管占了地球表面的71%，广阔的海洋仍然是一个几乎没有营养、极少有生命的巨大的水上沙漠。水生生物群落难以像在陆地上生物群落那样明确划分。可溶性营养素可控制许多地区水生生物的分布。水生生物群落可分为淡水（内陆的）生物群落和海水（咸水或海洋的）生物群落。

Unit 8

Text A

Amino Acid Analysis

At a low level of resolution, we can determine the amino acid composition of the protein by hydrolyzing the protein in 6 N HCl, 100°C, under vacuum for various time intervals. After removing the HCl, the hydrolyzate is applied to an ion-exchange or hydrophobic interaction column, and the amino acids eluted and quantitated with respect to known standards. A non naturally-occurring amino acid like nor-leucine is added in known amounts as an internal standard to monitor quantitative recovery during the reactions. The separated amino acids are often reacted with ninhydrin or phenylisothiocyantate to facilitate their detection. The reaction is usually allowed to procedure for 24, 36, and 48 hours, since amino acids with OH (like serine) are destroyed. A time course allows the concentration of Ser at time $t=0$ to be extrapolated. Trp is also destroyed during the process. In addition, the amide links in the side chains of Gln and Asn are hydrolyzed to form Glu and Asp, respectively.

N- and C-Terminal Amino Acid Analysis

The amino acid composition does not give the sequence of the protein. The N-terminus of the protein can be determined by reacting the protein with fluorodinitrobenzene (FDNB) or dansyl chloride, which reacts with any free amine in the protein, including the epsilon amino group of lysine. The amino group of the protein is linked to the aromatic ring of the DNB through an amine and to the dansyl group by a sulfonamide, and are hence stable to hydrolysis. The protein is hydrolyzed in 6 N HCl, and the amino acids separated by TLC or HPLC. Two spots should result if the protein was a single chain, with some Lys residues. The labeled amino acid other than Lys is the N-terminal amino acid. The C-terminal amino acid can be determined by addition of carboxypeptidases, enzymes which cleave amino acids from the C-terminal. A time course must be done to see which amino acid is released first. N-terminal analysis can also be done as part of sequencing the entire protein as discussed below (Edman degradation reaction).

Analysis for Specific Amino Acids

Aromatic amino acids can be detected by their characteristic absorbance profiles. Amino acids with specific functional groups can be determined by chemical reactions with specific modifying groups.

Amino Acid Sequence

Two methods exist to determine the entire sequence of a protein. In one, the protein is sequenced; in the other, the DNA encoding the protein is sequenced, from which the amino acid sequence can be derived. The actually protein can be sequenced by automated, sequential Edman Degradation. In this technique, a protein adsorbed to a solid phase reacts with phenylisothiocyanate. An intramolecular cyclization and cleavage of the N-terminal amino acid results, which can be washed from the adsorbed protein and detected by HPLC analysis. The yields in this technique are close to 100%. However, with time, more chains accumulate in which an N-terminal amino acid has not been removed. If it is removed on the next step, two amino acids will elute, creating increasing "noise" in the elution step—i. e. more than 1 amino acid derivative will be detected. Hence the maximal length of the peptide which can be sequenced is about 50 amino acids. Most proteins are larger than that. Hence, before the protein can be sequenced, it must be cleaved with specific enzymes called endoproteases which cleave proteins after specific side chains. For example, trypsin cleaves proteins within a chain after Lys and Arg, while chymotrypsin cleaves after aromatic amino acids, like Trp, Tyr, and Phe. Chemical cleavage by small molecules can be used as well. Cyanogen bromide, CNBr, cleaves proteins after methionine side chains. The individual proteins must be cleaved using two different methods, and each peptide fragment isolated and sequenced. Then the order of the cleaved peptides with known sequence can be pieced together by comparing the peptide sequences obtained using different cleavage methods. Many proteins also have disulfide bonds connecting Cys side chains distal to each other in the polypeptide chain. Proteolytic or chemical cleavage of the protein would lead to the formation of a fragment containing two peptides linked by disulfides. Edman degration would release two amino acids from such fragments. To avoid this problem, the protein is oxidized with performic acid, which irreversibly oxidizes free Cys, or Cys-Cys disulfides to cysteic acid residues.

New Words

analysis　　　　　　　　［əˈnæləsis］　　　　　　n. 分析
resolution　　　　　　　［ˌrezəˈljuːʃən］　　　　　n. 溶解，分解

hydrolyze	['haidrəlaiz]	vi.	水解
vacuum	['vækjuəm]	n.	真空
		adj.	真空的
interval	['intəvəl]	n.	间隔；距离
hydrolyzate	[hai'drɔlə‚zeit]	n.	水解产物
elute	[i'lju:t]	vt.	洗脱，洗提
quantitate	['kwɔntiteit]	v.	测定（估计）的数量，用数量来表示
norleucine	['nɔ:ljusi:n]	n.	己氨酸，正亮氨酸
ninhydrin	[nin'haidrin]	n.	茚三酮
serine	['seri:n]	n.	丝氨酸（＝Ser）
amide	['æmaid]	n.	酰胺
fluorodinitrobenzene	[‚flu:ərə‚dai‚naitrə'benzi:n]	n.	氟二硝基苯
aromatic	[‚ærəu'mætik]	adj.	芳香族的
sulfonamide	[sʌl'fɔnəmaid]	n.	磺酰胺
hydrolysis	[hai'drɔlisis]	n.	水解
carboxypeptidase	[ka:‚bɔksi'peptideis]	n.	羧肽酶
absorbance	[əb'sɔ:bəns]	n.	吸光度，吸光指数，吸光率
cyclization	[‚saikli'zeiʃən]	n.	环化
derivative	[di'rivətiv]	n.	衍生物
chymotrypsin	[‚kaimə'tripsin]	n.	胰凝乳蛋白酶，糜蛋白酶
methionine	[me'θaiəni:n]	n.	蛋氨酸，甲硫氨酸
peptide	['peptaid]	n.	肽，缩氨酸
proteolytic	[prəutiə'litik]	adj.	蛋白水解的
disulfide	[daisʌlfaid]	n.	二硫化物

Phrases

hydrophobic interaction column	疏水作用柱
with respect to	谈到，关于
quantitative recovery	回收的量
dansyl chloride	丹磺酰氯
degradation reaction	降解反应
cyanogen bromide	溴化氰

Notes

[1] The C-terminal amino acid can be determined by addition of carboxypeptidases, en-

zymes which cleave amino acids from the C-terminal.

本句中,enzymes which cleave amino acids from the C-terminal 是一个名词短语,对 carboxypeptidases 作进一步补充说明。在该短语中,which cleave amino acids from the C-terminal 是一个定语从句,修饰和限定 enzymes。

[2] An intramolecular cyclization and cleavage of the N-terminal amino acid results, which can be washed from the adsorbed protein and detected by HPLC analysis.

本句中,which can be washed from the adsorbed protein and detected by HPLC analysis, the N-terminal amino acid 是一个非限定性定语从句,对 the N-terminal amino acid 进行补充说明。

[3] Hence, before the protein can be sequenced, it must be cleaved with specific enzymes called endoproteases which cleave proteins after specific side chains.

本句中,是一个时间状语从句,called endoproteases 是一个过去分词短语,作定语,修饰和限定 specific enzymes。which cleave proteins after specific side chains 是一个定语从句,修饰和限定 endoproteases。

[4] To avoid this problem, the protein is oxidized with performic acid, which irreversibly oxidizes free Cys, or Cys-Cys disulfides to cysteic acid residues.

本句中,To avoid this problem 是一个动词不定式短语,作目的状语。which irreversibly oxidizes free Cys, or Cys-Cys disulfides to cysteic acid residues 是一个非限定性定语从句,对 performic acid 进行补充说明。

Exercises

【EX.1】 根据课文内容,回答以下问题

1) When can we determine the amino acid composition of the protein?

2) How can we determine the amino acid composition of the protein?

3) How can the N-terminus of the protein can be determined?

4) What is the amino group of the protein linked?

5) How can the C-terminal amino acid be determined?

6) How can amino acids with specific functional groups be determined?

7) What are the two methods to determine the entire sequence of a protein?

8) What is the maximal length of the peptide which can be sequenced?

9) Where does trypsin cleave proteins?

10) Where does chymotrypsin cleave proteins?

【EX. 2】 根据下面的英文解释,写出相应的英文词汇

英 文 解 释	词 汇
The act or process of separating or reducing something into its constituent parts.	
Decomposition of a chemical compound by reaction with water.	
To extract (one material) from another, usually by means of a solvent.	
To determine or measure the quantity of.	
A compound derived or obtained from another and containing essential elements of the parent substance.	
The amount of time between two specified instants, events, or states.	
A chemical compound containing two sulfur atoms combined with other elements or radicals.	
Any of various natural or synthetic compounds containing two or more amino acids linked by the carboxyl group of one amino acid and the amino group of another.	
Any of a group of organic sulfur compounds containing the radical ONH and including the sulfa drugs.	
The formation of one or more rings in a hydrocarbon.	

【EX. 3】 把下列句子翻译为中文

1) Matter is composed of atoms. Each atom consists of a positively charged nucleus of protons and neutrons, surrounded by electrons bearing negative charges. There are many elements in nature, but only a few of them make up the bulk of living systems.

2) Isotopes of an element differ in their numbers of neutrons. Some isotopes are radioactive, emitting radiation as they decay.

3) Covalent bonds are strong bonds formed when two atomic nuclei share one or more pairs of electrons. Covalent bonds have spatial orientations that give molecules three-dimensional shapes.

4) Nonpolar molecules interact very little with polar molecules, including water. Nonpolar molecules are attracted to one another by very weak bonds called van der Waals forces.

5) Functional groups make up part of a larger molecule and have particular chemical properties. The consistent chemical behavior of functional groups helps us understand the properties of the molecules that contain them.

6) Macromolecules have specific, characteristic three-dimensional shapes that depend on the structure, properties, and sequence of their monomers.

7) The functions of proteins include support, protection, catalysis, transport, defense, regulation, and movement. Protein function sometimes requires an attached prosthetic group.

8) There are 20 amino acids found in proteins. Each amino acid consists of an amino group, a carboxyl group, a hydrogen, and a side chain bonded to the carbon atom.

9) Amino acids are covalently bonded together into polypeptide chains by peptide linkages, which form by condensation reactions between the carboxyl and amino groups.

10) Although lipids can form gigantic structures, these aggregations are not chemically macromolecules because the individual units are not linked by covalent bonds.

【EX. 4】 把下列短文翻译为中文

Activation of Amino Acids

Activation of amino acids is carried out by a two-step process catalyzed by aminoacyl-tRNA synthetases. Each tRNA and the amino acid it carries, are recognized by individual aminoacyl-tRNA synthetases. This means there exists at least 20 different aminoacyl-tRNA synthetases, there are actually at least 21 since the initiator met-tRNA of both prokaryotes and eukaryotes is distinct from non-initiator met-tRNAs.

Activation of amino acids requires energy in the form of ATP and occurs in a two step reaction catalyzed by the aminoacyl-tRNA synthetases. First the enzyme attaches the amino acid to the α-phosphate of ATP with the concomitant release of pyrophosphate. This is termed an aminoacyl-adenylate intermediate. In the second step the enzyme catalyzes transfer of the amino acid to either the 2'-OH or 3'-OH of the ribose portion of the 3'-terminal adenosine residue of the tRNA generating the activated aminoacyl-tRNA. Although these reactions are freely reversible, the forward reaction is favored by the coupled hydrolysis of PPi.

Text B

Glycoprotein Biosynthesis and Function

1. Glycoprotein Biosynthesis

Carbohydrates are added to proteins in a very complicated process which involves two organelles, the endoplasmic reticulum and the Golgi apparatus. CHO addition to proteins occurs both co- and post-translationally. The RNA coding the protein sequence enters the cytoplasm where it binds to ribosomes (large RNA-protein complexes) which are the sites for protein synthesis. Cytoplasmic proteins are synthesized on free ribosomes, but for future glycoproteins, the ribosomes bind to an elongated, extensive organelle in the cell called the endoplasmic reticulum (ER). The nascent protein chain enters the lumen of the ER and a core oligosaccharide is added to the protein. Further additions and removal (trimming and processing) of monosaccharides are preformed in the lumen until a final core mannose structure has been added. The ER lumen contains high concentrations of molecular chaperones to assist protein folding.

Now additional carbohydrate modifications (post-translational) are made as the protein moves from the lumen of the ER (probably by a budding process) to another series of stacked, pancake-like organelles called the Golgi apparatus. Here terminal carbohydrate modification is completed. The Golgi does not contain molecular chaperons since protein folding is complete when the proteins arrive. Rather they have high concentrations of membrane bound enzymes, including glycosidases, and glycosyl transferases.

2. Glyprotein Function

The role of CHO in glycoprotein structure/function is slowly being determined. The most important seems to involve their role in directing proper folding of proteins in the ER which accounts for the observations that glycan addition to proteins in the ER is a cotranslational event. When inhibitors of ER glycosylation are added to cells, protein misfolding and aggregation are observed. The extent of misfolding depends on the particular protein and particular glycosylation sites with the protein. The polar CHO residues help promote solubility of folding intermediates, similar to the effects of many chaperone proteins.

The glycan moieties of the folding glycoprotein also lead to binding of the protein to lectins in the ER which serve as molecular chaperones. The most studied of these chaperones are involved in the calnexin-calreticulin cycle, and facilitate correct disulfide bond formation in the protein. After two glucose residues are removed by glucosidase I and II, the monoglucosylated protein binds to calnexin (CNX) and/or calreticulin (CRT), two homologous ER lectins specific for monoglucosylated proteins. Once bound, another protein, ERp57, a molecular chaperone with a disulfide bond, interacts with the protein. This protein has protein disulfide isomerase activity.

If a glycoprotein has not folded completely, it is recognized by a glycoprotein glucosyltransferase, which adds a glucose to it. This then promotes reentry into the calnexin/calreticulin cycle.

Ideally, unfolded or misfolded proteins would be targeted from degradation and elimination from cells. The ER has evolved a system to accomplish this. Since folding occurs in the ER, to prevent misfolding and aggregation, the ER also contains chaperones and folding catalysts. Stress (such as through heat shock) stimulates ER chaperone activity. As a final defense mechanism, unfolded or aberrantly-folded proteins are degraded by the cytoplasmic proteasome complex. Nonnative forms of some proteins that "escape" this surveillance system can accumulate and result in disease (for example neurodegenerative diseases like Alzheimers and Parkinson's disease).

3. Glycobiology

Our understanding of the synthesis and structure of glycan portions of glycoproteins has lagged behind our understanding of protein and nucleic acid structure and synthesis. Several reasons account for this:

- carbohydrates are much more complex with more functional groups per carbon and with a much larger number of stereocenters, making chemical synthesis and structure determination more difficult.
- carbohydrate chain synthesis is not directed by a template as is the synthesis of DNA, RNA, and proteins.
- synthesis is spread over two different organelles, and which allows great heterogeneity in main chain and branch chain synthesis, which provides heterogeneous samples for analysis.

New techniques in analysis and synthesis of carbohydrate analogs and inhibitors of enzymes involved in CHO synthesis and degradation, as well as in genetic manipulations of gene for these enzymes, is revolutionizing our understanding of the function of carbohydrate groups on lipids and proteins. On a more practical note, new methods to synthesize glycoproteins using recombinant DNA technology have been developed that allow synthesis of therapeutic human glycoproteins in yeast. Although they share many of the same synthetic steps, yeast glycoproteins are enriched in the high-mannose type, making them targets of the human immune system. Human and yeast glycoproteins synthesis produce the same mannose core in the ER. However, differences in synthesis occurs in the Golgi. Human Golgi contain mannosidases Ⅰ and Ⅱ, which remove all but 3 Man residues from the final product. In yeast, however, these mannosidases appear to be missing so more Man residues are added (as many as 100). Human proteins made in yeast, therefore, contain many Man residues, which are recognized by the human immune system. So address these issues, Hamilton et al produced mutants of the yeast Pichia pastoris that localized glycoprotein synthesis proteins for mannosidase Ⅰ and Ⅱ, as well as other human glycoprotein synthesis genes, to the correct intracellular location while inactivating normal yeast gene, resulting in the production of human glycoproteins with the correct CHO structure in yeast. This may prove to have widespread use in the production of therapeutic human proteins.

New Words

carbohydrate	[ˈkɑːbəuˈhaidreit]	n. 碳水化合物
cytoplasm	[ˈsaitəuplæzm]	n. 细胞质
nascent	[ˈnæsnt]	adj. 初生的
oligosaccharide	[ˌɔligəuˈsækəraid]	n. 低聚糖,寡糖
monosaccharide	[ˌmɔnəuˈsækəraid]	n. 单糖
mannose	[ˈmænəus]	n. 甘露糖
glycan	[ˈglaikæn]	n. 多糖,聚糖,多聚糖
glycosidase	[glaiˈkəusideis]	n. 糖苷酶
glycosyl	[ˈglaikəusil]	n. 糖基
transferase	[ˈtrænsfəˌreis]	n. 转移酶
glycosylation	[ˌglaikəsiˈleiʃən]	n. 糖基化
solubility	[ˌsɔljuˈbiliti]	n. 溶解度,可溶性
lectin	[ˈlektin]	n. 凝集素,植物血凝素
chaperone	[ˈʃæpərəun]	n. 伴侣
glucosidase	[gluːˈkəusideis]	n. 葡萄糖苷酶
isomerase	[aiˈsɔməreis]	n. 异构酶
glucosyltransferase	[ˌgluːkəsilˈtrænsfəreis]	n. 葡萄糖基转移酶
aggregation	[ˌægriˈgeiʃən]	n. 聚集;集合
stereocenter	[ˈstiəriəˌsentə]	n. 立构中心
heterogeneity	[ˌhetərəudʒiˈniːiti]	n. 异质性

yeast	[ji:st]	n.	酵母
mannosidase	['mænəsideis]	n.	甘露糖苷酶

Phrases

Golgi apparatus	高尔基体
cytoplasmic protein	胞浆蛋白
free ribosome	游离核糖体
endoplasmic reticulum	内质网
membrane bound enzyme	膜结合酶
account for	说明，解释
calnexin-calreticulin cycle	钙连蛋白-钙网蛋白循环

Exercises

【EX. 5】 根据课文内容，回答以下问题

1) What are the two organelles involved when carbohydrates are added to proteins?

2) What are the sites for protein synthesis?

3) What does the ER lumen contain?

4) Why doesn't the Golgi contain molecular chaperons?

5) When are protein misfolding and aggregation observed?

6) What happens after two glucose residues are removed by glucosidase Ⅰ and Ⅱ?

7) What happens if a glycoprotein has not folded completely?

8) Which do we understand better, the synthesis and structure of glycan portions of glycoproteins or protein and nucleic acid structure and synthesis?

9) What do human and yeast glycoproteins synthesis produce?

10) What do human proteins made in yeast contain?

Reading Material

Text	Notes
Hydrogen Bonding[1] Linus Pauling first suggested that H bonds (between water and the protein and within the protein itself) would play a dominant role in protein folding and stability. It would seem to make sense[2] since amino acids are dipolar and secondary structure is common. Remember, however, the H bonds would be found not only in the native state but also in the denatured[3] state. Do they contribute differently to the stability of the D vs N states? Many experimental and theoretical studies have been performed investigating helix ⇌ (random) coil transitions in small peptides. Remember all the intrachain[4] H bonds in the helix[5]? Are they collectively more stable than H bonds between water and the peptide in a (random) coil? Early models assumed that intrachain H bonds were energetically more favorable than H bonds between peptide and water. But to form an H bond requires an entropy payback since a helix is much more ordered (lower entropy) than a random coil (higher entropy). At low temperature, enthalpy[6] predominates and helix formation in solution is favored. At high temperature, the helix is disfavored entropically. Imagine the increased vibrational and rotational states permitted to the atoms at higher temperatures. (Remember the trans to gauche conformational changes in the acyl chains of double chain amphiphiles as the temperature increased, leading to a transition from a gel to liquid crystalline phase in bilayer vesicles.) Theoretical studies on helix-coil transitions predict the following: • as the chain length increases, the helix gets more stable; • increasing the charge on the molecule destabilizes the helix, since the coil, compared to the more compact helix, has a lower charge density; • solvents that protonate[7] the carbonyl oxygen (like formic acid) destabilizes the helix; and • solvents that form strong H bonds compete with the peptide	[1]氢键 [2]有意义,有道理 [3]变性的 [4]链内的 [5]螺旋 [6]焓 [7]加质子于……

Text	Notes
and destabilize the helix. In contrast, solvents such as CHCl3, dimethylformamide (a nonprotic solvent), or 2-chloroethanol, and trifluoroethanol, which form none or weaker H bonds to the peptide than does water, stabilizes the helix.	
These helix-coil studies suggest that H bonds are important in stabilizing a protein.	
But do they really? Why should these H bonds differ from those in water? It's difficult to figure out whether they are since there are so many possible H bonds (between protein and water, water and water, and protein and protein), and their strength depends on their orientation and the dielectric[8] constant of the medium in which they are located.	[8]电介质，绝缘体
If intrachain H bonds in a protein are not that much different in energy than intermolecular H bonds between the protein and water, and given that proteins are marginally[9] stable at physiological temperatures, then it follows that the folded state must contain about as many intramolecular hydrogen bonds within the protein as possible intermolecular H bonds between the protein and water, otherwise the protein would unfold.	[9]在边上
To resolve this issue, and determine the relative strength of H bonds between the varying possible donors[10] and acceptors[11], many studies have been conducted to compare the energy of H bonds between small molecules in water with the energy of H bonds between the same small molecules but in a nonpolar solvent. The rationale goes like this. If the interior of a protein is more nonpolar than water (lower dielectric constant than water), then intrastrand H bonds in a protein might be modeled by looking at the H bonds between small molecules in nonpolar solvents and asking the question, is the free energy change for the following process < 0:	[10]供体 [11]受体
$$Dw + Aw \rightleftharpoons (DA)n, \triangle G°, K$$	
where D is a hydrogen bond donor (like NH) and A is a hydrogen bond acceptor, (like C=O), w is water (i.e. donor and acceptor are in water), and n is a nonpolar solvent, and $\triangle G°$ and K are the standard free energy change and the equilibrium constant, respectively, for the formation of a H-bond in a nonpolar solvent from a donor	

Text	Notes
and acceptor in water. This reaction simulates H-bond contributions to protein folding, where a buried H-bond is mimicked[12] by a H-bond in a nonpolar solvent. The reaction written above is really a thought experiment, since it would be hard to set up the necessary conditions to make the measurement. However, we can calculate the $\triangle G°$ for this reaction since it is a state function and it really doesn't matter how one accomplishes this process.	[12]模仿

Text A 参考译文

氨基酸分析

溶解度较低时,我们可通过在真空状态下,在6mol/L HCl、100℃的溶液中,以不同时间间隔来水解蛋白质,以确定氨基酸组成成分。除去HCl之后,水解产物过离子交换或疏水作用柱,用已知标准样洗脱和定量测定氨基酸。添加已知数量的非天然存在氨基酸(如已氨酸)作为一种内标物来监测反应中回收的量。分离的氨基酸常与茚三酮或苯异硫氰酸酯作用,这样易于测定。由于要破坏具有羟基的氨基酸(如丝氨酸),这种反应通常要持续24、36或48h。通过这个时长可推测出丝氨酸的浓度。色氨酸在这个处理过程中也被破坏。另外,谷氨酰胺和天冬酰胺侧链中的酰胺键被水解,分别形成了谷氨酸和天冬氨酸。

N-末端和C-末端氨基酸分析

从氨基酸的含量不能给出蛋白质的序列。蛋白质与氟二硝基苯(FDNB)或丹磺酰氯发生反应可测定蛋白质N末端氨基酸,这两种物质可与蛋白质中任一游离氨基发生反应,包括赖氨酸的ε-氨基。蛋白质氨基通过一个胺被连接到二硝基苯的芳香环上,而且通过磺酰胺被连接到丹磺酰基,因而具有水解稳定性。蛋白质在6mol/L HCl溶液中水解,再通过薄层色谱法(TLC)或高效液相色谱法(HPLC)将氨基酸分离。如果蛋白质是单链,那么就会导致两个位点的产生,同时伴随着一些赖氨酸残基的出现。已标记的氨基酸除了赖氨酸外都是N末端氨基酸。C末端氨基酸可通过外加羧肽酶——一种从C末端裂解氨基酸的酶来测定。要知道哪一种氨基酸最先被释出需要一个时长时间。N末端分析也可作为下述的整个蛋白质测序的一部分(Edman降解反应)。

特殊氨基酸分析

芳香族氨基酸的检测,可利用其特有的吸光性。具有特殊官能团的氨基酸与特殊修饰基团发生化学反应,这样就可以测定这些氨基酸。

氨基酸序列

现有两种方法用于测定整个蛋白质的序列。第一种方法是直接对蛋白质进行测序；另一种方法是先对编码蛋白质的 DNA 进行测序，再推导出氨基酸的序列。实际上蛋白质的测序是由自动连续的 Edman 降解法来完成的。这种方法中，被吸附到固相的蛋白质与苯异硫氰酸酯起反应，导致分子内环化和 N 末端氨基酸的裂解，N 末端氨基酸可从被吸附的蛋白质洗出，再用 HPLC 分析法测得。本方法氨基酸产出接近 100%。然而，随着时间的延长，由于 N 末端氨基酸没有被去除，就会累积更多氨基酸链。如果 N 末端氨基酸在接下来的步骤中被去除，那么两个氨基酸将会被洗脱，这将在洗脱步骤中产生更多的"噪音"，也就是说，测得不止一个的氨基酸衍生物。因此，可被测序的肽链最长大约是 50 个氨基酸长度。大多数蛋白质的氨基酸数是大于这个数的，所以，在蛋白质测序之前，必须首先用一种称作蛋白内切酶的特殊酶来切开，这种酶是在特异侧链之后切开蛋白质。例如，胰岛素在一条链中的赖氨酸和精氨酸之后切开蛋白质，而糜蛋白酶在芳香族氨基酸，如色氨酸、酪氨酸和苯丙氨酸之后切开蛋白质。也可采用小分子的化学切割。如溴化氰和 CNBr，在侧链的蛋氨酸之后切开蛋白质。单个蛋白质必须用两种不同的方法来切开，每一个肽片段都被分离和测序。对于已被切开且已知序列的肽，可通过比较采用不同的切开方法获得的肽序列拼凑出来它的序列顺序。许多蛋白质中也有连接多肽链中的两个半胱氨酸侧链末端的二硫键。蛋白质的水解切割或化学切割可能会导致由二硫化物连接在一起的二肽片段的形成。Edman 降解将会从这样的片段中释放出两个氨基酸。为了防止这个问题的出现，蛋白质被过甲酸氧化，过甲酸能使游离的半胱氨酸或半胱氨酸-半胱氨酸二硫化物发生不可逆氧化，生成半胱氨酸残基。

Text B 参考译文

糖蛋白生物合成和功能

1. 糖蛋白生物合成

碳水化合物经一个非常复杂的过程被结合到蛋白质中，这一复杂过程与两个细胞器——内质网和高尔基体有关。碳水化合物附加到蛋白质中，不仅发生在翻译同步过程中，而且也发生在翻译后过程中。对蛋白质序列进行编码的 RNA 进入细胞质，在那儿与核糖体结合在一起（大的 RNA-蛋白质复合体），核糖体是蛋白质的合成场所。胞浆蛋白在游离核糖体上合成，但为了合成糖蛋白，核糖体结合到细胞内一种伸长的、分布范围广被称为内质网（ER）的细胞器上。新生成的蛋白质进入内质网内腔，并有一种核心低聚糖被添加到这个蛋白质中。接着，单糖在内腔进行进一步的添加与切除（修饰和加工），直至最后一个核心甘露糖结构被添加到蛋白质上。内质网内腔包含有高浓度的分子伴侣来帮助蛋白质折叠。

当蛋白质从内质网（ER）的内腔（经由一个出芽过程）转移到一个叠层状、薄饼样的高尔基体时，就发生了新的额外的碳水化合物修饰（翻译后），此时最终的碳水化合物的修饰就完成了。当蛋白质到达高尔基体时，由于蛋白质进来时的折叠已经完成，故高尔基体不含分子伴侣，取而代之的是高浓度的膜结合酶，包括糖苷酶和糖基转移酶。

2. 糖蛋白的功能

碳水化合物在糖蛋白的结构和功能上的作用正在逐渐被揭示。它最重要的作用是引导内质网中蛋白质进行正确折叠,这说明了一个现象,即在内质网内多糖添加到蛋白质中是一个协同翻译过程。当细胞添加内质网糖基化作用抑制剂时,就能观察到蛋白质的错误折叠和凝集。错误折叠的程度取决于特异的蛋白质和特异糖基化作用在蛋白质中位点。极性 CHO 残基有助于提高折叠中介物的溶解度,折叠中介物的作用与许多伴侣蛋白的作用相似。

折叠糖蛋白的多糖部分也会引起蛋白质黏合到内质网内作为分子伴侣的凝集素上。其中研究的最多的是与钙连蛋白-钙网蛋白循环和促进蛋白质中正确的二硫键形成有关的分子伴侣。在两个葡萄糖残基被葡萄糖苷酶Ⅰ和Ⅱ去除之后,单葡糖苷酶催化的蛋白质结合到两种对单葡糖苷酶催化的蛋白质具有特异性的钙连蛋白和钙网蛋白上。一旦结合,另一种具有一个二硫键的分子伴侣 ERp57 蛋白质可与这个蛋白质结合。这个蛋白质就有了二硫键的构象活性。

如果一种糖蛋白没有折叠完全,就会被糖蛋白葡萄糖基转移酶识别,它能添加一些葡萄糖分子,然后增加进入钙连蛋白-钙网蛋白循环的次数。

理想状态下,不折叠和折叠错误的蛋白质将会成为细胞中降解和移除的目标。内质网已进化出一套系统来完成这个任务。既然折叠发生在内质网中,为防止折叠错误和凝集,内质网中也应具有分子伴侣和折叠的催化剂。刺激(如通过热休克)激发内质网中分子伴侣的活性。作为一种最终的防御机制,不折叠和异常折叠的蛋白质将由细胞质中蛋白酶复合体降解。非天然形成"逃过"监测系统的一些蛋白质会在体内累积引起疾病(例如神经变性疾病,包括阿尔茨海默病和帕金森病)。

3. 糖生物学

对于糖蛋白中多聚糖部分的合成和结构的认识,落后于人们对于蛋白质和核酸结构和合成的认识。原因有以下几点:

- 碳水化合物具有每一个碳中有功能更强的基团和更多数量的立构中心,因此更为复杂,这也使得要确定它们的化学合成和结构就更为困难。
- 碳水化合物的合成不像 DNA、RNA 和蛋白质的合成那样直接有一个模板。
- 合成反应是在两个不同的细胞器中进行的,这就使得碳水化合物的主链和侧链的合成都具有很大的异质性,也使得它们可为分析提供不同的样本。

关于碳水化合物的合成与降解的酶的抑制剂和碳水化合物类似物的分析和合成以及这些酶的基因操作,现在有了一些新技术。这些新技术极大地加深了人们对结合在脂类和蛋白质上的碳水化合物基团的理解。在一份更实用的文献中记录了一种利用 DNA 重组技术来合成糖蛋白的新方法,利用这种新方法能在酵母中合成治疗用的人类糖蛋白。尽管有很多的合成步骤是相同的,但酵母糖蛋白中富含高甘露糖,这可使人类对这种糖蛋白产生免疫反应。人类和酵母糖蛋白的合成在内质网内产生同样的核心甘露糖,但是合成的差异发生在高尔基体内。人类的高尔基体含有甘露糖苷酶

Ⅰ和Ⅱ,它们能从最终产物中移去除3个人类残基外的其他残基。但在酵母中,似乎不存在这些甘露糖苷酶,所以甘露糖残基可被添加100个之多。因此,由酵母产生的人类蛋白质因此含有更多的人类残基,被人类免疫系统识别。所以,针对这个问题,Hamilton等人获得了酵母 *Pichia pastoris* 的突变种,可以将糖苷酶Ⅰ和Ⅱ和其他人类糖蛋白合成基因的糖蛋白合成蛋白定位到正确的细胞位点,同时抑制正常的酵母基因,结果在酵母中会产生具有正确的CHO结构的人类糖蛋白。在生产治疗性的人类蛋白的过程中,这种方法可能会被证明具有广泛的应用前景。

Unit 9

Text A

Immune Response

Among the many threats organisms face are invasion and infection by bacteria, viruses, fungi, and other foreign or disease-causing agents. All organisms have nonspecific defenses (or innate defenses) that provide them with some of the protection they need. This type of defense exists throughout the animal kingdom, from sponges to mammals. Vertebrate animals, however, have an additional line of defense called specific immunity. Specific immunity is also called acquired immunity, adaptive immunity, or, most simply, an immune response.

Overview

One characteristic of specific immunity is recognition. Immune responses begin when the body recognizes the invader as foreign. This occurs because there are molecules on foreign cells that are different from molecules on the body's cells. Molecules that start immune responses are called antigens. The body does not usually start an immune response against its own antigens because cells that recognize self-antigens are deleted or inactivated. This concept is called self-tolerance and is a key characteristic that defines immune responses.

A second characteristic is specificity. Although all immune responses are similar, each time the body is invaded by a different antigen, the exact response is specific to that antigen. For example, infection with a virus that causes the common cold triggers a response by a different set of cells than infection with bacteria that causes strep throat. A third characteristic is memory. After an antigen is cleared from the body, immunological memory allows an antigen to be recognized and removed more quickly if encountered again.

Antigen Presentation

Three groups of white blood cells are involved in starting an immune response.

Although immune responses can occur anywhere in the body these cells are found, they primarily occur in the lymph nodes and spleen. These organs contain large numbers of antigen-presenting cells (APCs), T lymphocytes (or T cells), and B lymphocytes (or B cells).

APCs include macrophages, dendritic cells, and B cells. These cells encounter the foreign invader and present the invader's antigens to a group of T cells called helper T cells (TH cells). APCs engulf an invader and bring it inside the cell. The APC then breaks the invader apart into its antigens and moves these antigens to its cell surface. Receptors are cell surface proteins that can attach to antigens. Each TH cell has a different receptor, allowing each cell to recognize a different antigen.

The APC "shows" the antigen to the TH cells until there is a match between a TH cell receptor and the antigen. The contact between the two cells stimulates the TH cell to divide rapidly. This process is called clonal selection because only the TH cells that recognize the foreign invader are selected to reproduce. Stimulated TH cells also produce chemical messengers called cytokines. Cytokines are made by all immune cells and control the immune response.

Antigen Clearance

The large numbers of TH cells activate two other populations of white blood cells: cytotoxic T cells (TC cells) and B cells. Like TH cells, each TC cell and B cell has receptors that match one antigen. This is why the immune system can recognize millions of antigens with specificity. The cells with the appropriate receptor encounter the antigen, preparing them for activation. They receive the final signal necessary for clonal selection from TH cells and cytokines. Cloned TC cells attach to invaders they recognize and release a variety of chemicals that destroy the foreign cell. Because this must happen through cell-to-cell contact, it is called cell-mediated immunity (or cellular immunity). It is especially effective at destroying abnormal body cells, such as cancerous cells or virus-infected cells. Cloned B cells destroy foreign invaders differently. After activation by TH cells, B cells release proteins called antibodies. Antibodies travel through the body's fluids and attach to antigens, targeting them for destruction by nonspecific defenses. This type of immune response is called antibodymediated immunity (or humoral immunity). It is especially effective at destroying bacteria, extracellular viruses, and other antigens found in body fluids.

Immunologic Memory

A primary immune response happens the first time that the body encounters a specific antigen. It takes several days to begin and one or two weeks to reach maximum activity. A secondary immune response occurs if the body encounters the same antigen at a later time. It takes only hours to begin and may peak within a few days. The invader is usually removed before it has a chance to cause disease. This is because some of the cloned TC cells and B cells produced during a primary immune response develop into memory cells. These cells immediately become activated if the antigen appears again. The complex interactions among cells described above are not necessary.

In fact, this is what happens when an individual is immunized against a disease. The vaccination (using weakened or killed pathogens) causes a primary immune response (but not the disease) and the production of memory cells that will provide protection if exposed to the disease-causing agent.

Immune System Disorders

Studying immune responses also allows scientists to understand immune system diseases. For example, hypersensitivity disorders occur when the immune system overreacts to an antigen, causing damage to healthy tissues. The result of this excessive antibody and TC cell activity can be relatively harmless (as with allergies to pollen, poison ivy, or molds) or deadly (as with autoimmune diseases or allergies to bee venom and antibiotics). At the opposite end of the spectrum are immunodeficiency diseases, conditions in which the body does not respond effectively against foreign invaders. HIV (human immunodeficiency virus) infection causes AIDS (acquired immunodeficiency syndrome) by attacking TH cells. Occasionally an individual is born with a deficient immune system, but these disorders are usually acquired (for example, from radiation treatment, chemotherapy, or infection with HIV). Whatever the cause, the individual has a more difficult time fighting infections.

New Words

immune	[iˈmjuːn]	adj.	免疫的
agent	[ˈeidʒənt]	n.	因子
invader	[inˈveidə]	n.	入侵物,入侵者
invade	[inˈveid]	vt.	侵略,侵袭
strep	[strep]	n.	链(锁状)球菌
specificity	[ˌspesiˈfisiti]	n.	特异性
inactivate	[inˈæktiveit]	vt.	使不活动;使变不活泼;去激活
spleen	[spliːn]	n.	脾
lymphocyte	[ˈlimfəsait]	n.	淋巴球,淋巴细胞
macrophage	[ˈmækrəfeidʒ]	n.	巨噬细胞
engulf	[inˈgʌlf]	vt.	吞没;吞食(=ingulf)

dendritic	[den'dritik]	adj. 树枝状的
cytokine	['saitəu'kain]	n. 细胞因子,细胞活素类物质
cytotoxic	[ˌsaitə'tɔksik]	adj. 细胞毒素的
activation	[ˌækti'veiʃən]	n. 活化,激活
vaccination	[ˌvæksi'neiʃən]	n. 接种疫苗
pathogens	['pæθədʒəns]	n. 病原体
hypersensitivity	[ˌhaipəsensi'tiviti]	n. 超敏性
overreact	[ˌəuvəri'ækt]	vi. 反应过度
allergy	['ælədʒi]	n. 过敏
venom	['venəm]	n. 毒液
immunodeficiency	[ˌimjunəudi'fiʃənsi]	n. 免疫缺陷
autoimmune	[ˌɔːtəvi'mjuːn]	adj. 自体免疫的
chemotherapy	[ˌkeməu'θerəpi]	n. 化学疗法,化疗

Phrases

immune response	免疫反映
immune system	免疫系统
nonspecific defense	非特异性防疫系统
specific immunity	特异性免疫
acquired immunity	获得性免疫
adaptive immunity	适应性免疫
strep throat	脓毒性咽喉炎
antigen presentation	抗原呈递
lymph node	淋巴结
clonal selection	克隆选择
antigen clearance	抗原清除
immunologic memory	免疫记忆
humoral immunity	体液免疫
bee venom	蜂毒

Abbreviations

APCs (antigen-presenting cells)	抗原呈递细胞
AIDS (acquired immunodeficiency syndrome)	艾滋病,获得性免疫缺陷综合征

Notes

[1] This occurs because there are molecules on foreign cells that are different from molecules on the body's cells.

本句中,because there are molecules on foreign cells that are different from molecules on the body's cells 是一个原因状语从句,修饰谓语 occurs。在该状语从句中,that are different from molecules on the body's cells 是一个定语从句,修饰和限定第一个 molecules。

[2] Although all immune responses are similar, each time the body is invaded by a different antigen, the exact response is specific to that antigen.

本句中,Although all immune responses are similar 是一个让步状语从句,each time the body is invaded by a different antigen 是一个时间状语从句,它们都修饰谓语 is specific to that antigen。

[3] Cloned TC cells attach to invaders they recognize and release a variety of chemicals that destroy the foreign cell.

本句中,they recognize 是一个省略了 that 的定语从句,修饰和限定 invaders。that destroy the foreign cell 也是一个定语从句,修饰和限定 a variety of chemicals。

[4] At the opposite end of the spectrum are immunodeficiency diseases, conditions in which the body does not respond effectively against foreign invaders.

为了保持句子的平衡,本句使用了倒装句。本句中,主语是 immunodeficiency diseases,conditions in which the body does not respond effectively against foreign invaders 是对 immunodeficiency diseases 的补充说明,in which the body does not respond effectively against foreign invaders 是一个定语从句,修饰和限定 conditions。are 是系动词,与 At the opposite end of the spectrum 一起作谓语。

Exercises

【EX. 1】 根据课文内容,回答以下问题

1) What are the threats mentioned in the text organisms face?

2) What are the three characteristics of immune response?

3) What are molecules that start immune responses called?

4) Why doesn't the body usually start an immune response against its own antigens?

5) After an antigen is cleared from the body, if encountered again, what does immunological memory do?

6) What do APCs include?

7) What are cytokines made? And what do they do?

8) What does antibodymediated immunity refer to?

9) The invader is usually removed before it has a chance to cause disease. Why?

10) What does HIV infection cause, and how?

【EX. 2】 根据下面的英文解释,写出相应的英文词汇

英 文 解 释	词 汇
Protected against a particular disease by particular substances in the blood.	
Of, relating to, or producing a toxic effect on cells.	
An innate, acquired, or induced inability to develop a normal immune response.	
A large, highly vascular lymphoid organ, lying in the human body to the left of the stomach below the diaphragm, serving to store blood, disintegrate old blood cells, filter foreign substances from the blood, and produce lymphocytes.	
Make sth. not work or operate any more, make inactive.	
To put or force in inappropriately, especially without invitation, fitness, or permission.	
The chemical messengers which are made by all immune cells and control the immune response.	
The treatment of disease using chemical agents or drugs that are selectively toxic to the causative agent of the disease, such as a virus, bacterium, or other microorganism.	
A poisonous secretion of an animal, such as a snake, spider, or scorpion, usually transmitted by a bite or sting.	
An abnormally high sensitivity to certain substances, such as pollens, foods, or microorganisms.	

【EX. 3】 把下列句子翻译为中文

1) The surface defenses of the body consist of the skin and the mucous membranes lining the digestive and respiratory tracts, which eliminate many microorganisms before they can invade the body tissues.

2) Inflammation aids the fight against infection by increasing blood flow to the site and raising temperature to retard bacterial growth.

3) The inflammatory response aids the mobilization of defensive cells at infected sites.

4) Nonspecific defenses include physical barriers such as the skin, phagocytic cells, killer cells, and complement proteins.

5) The immune system evolved in animals from a strictly nonspecific immune response in invertebrates to the two-part immune defense found in mammals.

6) Skin not only defends the body by providing a nearly impenetrable barrier, but also reinforces this defense with chemical weapons on the surface.

7) Physical and anatomic barriers that tend to prevent the entry of pathogens are an organism's first line of defense against infection.

8) Innate immunity is not specific to any one pathogen but rather constitutes a first line of defense, which includes anatomic, physiologic, endocytic and phagocytic, and inflammatory barriers.

9) Adaptive immune responses exhibit four immunologic attributes: specificity, diversity, memory, and self/nonself recognition.

10) Immunity is the state of protection against foreign organisms or substances (antigens). Vertebrates have two types of immunity, innate and adaptive.

【EX. 4】 把下列短文翻译为中文

Adaptive immunity responds to the challenge with a high degree of specificity as well as the remarkable property of "memory". Typically, there is an adaptive immune response against an antigen within five or six days after the initial exposure to that antigen. Exposure to the same antigen some time in the future results in a memory response: the immune response to the second challenge occurs more quickly than the first, is stronger, and is often more effective in neutralizing and clearing the pathogen. The major agents of adaptive immunity are lymphocytes and the antibodies and other molecules they produce.

Because adaptive immune responses require some time to marshal, innate immunity provides the first line of defense during the critical period just after the host's exposure to a pathogen. In general, most of the microorganisms encountered by a healthy individual are readily cleared within a few days by defense mechanisms of the innate immune system before they activate the adaptive immune system.

Text B

Cellular Counterattack: The Second Line of Defense

The surface defenses of the vertebrate body are very effective but are occasionally breached, allowing invaders to enter the body. At this point, the body uses a host of nonspecific cellular and chemical devices to defend itself. We refer to this as the second line of defense. These devices all have one property in common: they respond to any microbial infection without pausing to determine the invader's identity.

Although these cells and chemicals of the nonspecific immune response roam through the body, there is a central location for the collection and distribution of the cells of the immune system; it is called the lymphatic system. The lymphatic system consists of a network of lymphatic capillaries, ducts, nodes and lymphatic organs, and although it has other functions involved with circulation, it also stores cells and other agents used in the immune response. These cells are distributed throughout the body to fight infections, and also stored in the lymph nodes where foreign invaders can be eliminated as body fluids pass through.

Cells That Kill Invading Microbes

Perhaps the most important of the vertebrate body's nonspecific defenses are white blood cells called leukocytes that circulate through the body and attack invading microbes within tissues. There are three basic kinds of these cells, and each kills invading microorganisms differently.

Macrophages ("big eaters") are large, irregularly shaped cells that kill microbes by ingesting them through phagocytosis, much as an amoeba ingests a food particle. Within the macrophage, the membrane-bound vacuole containing the bacterium fuses with a lysosome. Fusion activates lysosomal enzymes that kill the microbe by liberating large quantities of oxygen free-radicals. Macrophages also engulf viruses, cellular debris, and dust particles in the lungs. Macrophages circulate continuously in the extracellular fluid, and their phagocytic actions supplement those of the specialized phagocytic cells that are part of the structure of the liver, spleen, and bone marrow. In response to an infection, monocytes (an undifferentiated leukocyte) found in the blood squeeze through capillaries to enter the connective tissues. There, at the site of the infection, the monocytes are transformed into additional macrophages.

Neutrophils are leukocytes that, like macrophages, ingest and kill bacteria by phagocytosis. In addition, neutrophils release chemicals (some of which are identical to household bleach) that kill other bacteria in the neighborhood as well as neutrophils themselves.

Natural killer cells do not attack invading microbes directly. Instead, they kill cells of the body that have been infected with viruses. They kill not by phagocytosis, but rather by creating a hole in the plasma membrane of the target cell. Proteins, called perforins, are released from the natural killer cells and inserted into the membrane of the target cell, forming a pore. This pore allows water to rush into the target cell, which then swells and bursts. Natural killer cells also attack cancer cells, often before the cancer cells have had a chance to develop into a detectable tumor. The vigilant surveillance by natural killer cells is one of the body's most potent defenses against cancer.

Proteins That Kill Invading Microbes

The cellular defenses of vertebrates are enhanced by a very effective chemical defense called the complement system. This system consists of approximately 20 different proteins that circulate freely in the blood plasma. When they encounter a bacterial or fungal cell wall, these proteins aggregate to form a membrane attack complex that inserts itself into the foreign cell's plasma membrane, forming a pore like that produced by natural killer cells. Water enters the foreign cell through this pore, causing the cell to swell and burst. Aggregation of the complement proteins is also triggered by the binding of antibodies to invading microbes.

The proteins of the complement system can augment the effects of other body defenses. Some amplify the inflammatory response by stimulating histamine release; others attract phagocytes to the area of infection; and still others coat invading microbes, roughening the microbes' surfaces so that phagocytes may attach to them more readily.

Another class of proteins that play a key role in body defense are interferons. There are three major categories of interferons: alpha, beta, and gamma. Almost all cells in the body make alpha and beta interferons. These polypeptides act as messengers that protect normal cells in the vicinity of infected cells from becoming infected. Though viruses are still able to penetrate the neighboring cells, the alpha and beta interferons prevent viral replication and protein assembly in these cells. Gamma interferon is produced only by particular lymphocytes and natural killer cells. The secretion of gamma interferon by these cells is part of the immunological defense against infection and cancer.

New Words

counterattack	[ˈkauntərəˌtæk]	n. 反击,反攻
breach	[briːtʃ]	vt. 打破,突破
property	[ˈprɔpəti]	n. 性质,特性
microbial	[maikrəubiəl]	adj. 微生物的
identity	[aiˈdentiti]	n. 身份;特性
roam	[rəum]	v. 漫游,闲逛,徘徊
location	[ləuˈkeiʃən]	n. 位置,场所,特定区域
duct	[dʌkt]	n. 管;输送管
microbe	[ˈmaikrəub]	n. 微生物,细菌
phagocytosis	[ˌfæɡəˌsaiˈtəusis]	n. 噬菌作用
amoeba	[əˈmiːbə]	n. 阿米巴,变形虫
vacuole	[ˈvækjuəul]	n. 空泡;液泡
lysosome	[ˈlaisəsəum]	n. 溶酶体
squeeze	[skwiːz]	n. 挤
neutrophil	[ˈnjuːtrəufil]	n. 嗜中性白细胞
vigilant	[ˈvidʒilənt]	adj. 警惕的;警觉的
surveillance	[səˈveiləns]	n. 监视,监督
aggregate	[ˈæɡriɡeit]	v. 聚集,集合
aggregation	[ˌæɡriˈɡeiʃ(ə)n]	n. 聚集,集合
phagocyte	[ˈfæɡəusait]	n. 食菌细胞;吞噬细胞
augment	[ɔːɡˈment]	v. 增加,增大
amplify	[ˈæmplifai]	vt. 放大,增强
histamine	[ˈhistəmiːn]	n. 组胺
roughen	[ˈrʌfən]	v. (使)变粗糙
interferon	[ˌintəˈfiərən]	n. 干扰素
polypeptide	[ˌpɔliˈpeptaid]	n. 多肽,缩多氨酸
penetrate	[ˈpenitreit]	vt. 穿透,渗透

Phrases

line of defense	防线

a host of	许多,一大群,大量
lymphatic system	淋巴系统
lymphatic capillary	淋巴毛细管
lymph node	淋巴结,淋巴结点
food particle	食物颗粒
lysosomal enzyme	溶解酶
dust particle	灰尘颗粒
bone marrow	骨髓
natural killer cell	自然杀伤细胞
complement system	补体系统
inflammatory response	炎症反应
in the vicinity of	在……附近;邻近……

Exercises

【EX. 5】 根据课文内容,回答以下问题

1) What is the second line of defense?

2) What property do nonspecific cellular and chemical devices have in common?

3) What is the lymphatic system?

4) What are the most important of the vertebrate body's nonspecific defenses? What do they do?

5) What are macrophages?

6) What are eutrophils?

7) What do natural killer cells do?

8) How are the cellular defenses of vertebrates enhanced?

9) What can the proteins of the complement system do?

10) How many major categories of interferons are there? What are they?

Reading Material

Text	Notes
The Immune Response: The Third Line of Defense 　　Few of us pass through childhood without contracting[1] some sort of infection. Chicken pox[2], for example, is an illness that many of us experience before we reach our teens. It is a disease of childhood, because most of us contract it as children and never catch it again. Once you have had the disease, you are usually immune to it. Specific immune defense mechanisms provide this immunity. Discovery of the Immune Response 　　In 1796, an English country doctor named Edward Jenner carried out an experiment that marks the beginning of the study of immunology[3]. Smallpox[4] was a common and deadly disease in those days. Jenner observed, however, that milkmaids who had caught a much milder form of "the pox" called cowpox[5] (presumably from cows) rarely caught smallpox. Jenner set out to test the idea that cowpox conferred protection against smallpox. He infected people with cowpox, and as he had predicted, many of them became immune to smallpox. 　　We now know that smallpox and cowpox are caused by two different viruses with similar surfaces. Jenner's patients who were injected with the cowpox virus mounted a defense that was also effective against a later infection of the smallpox virus. Jenner's procedure of injecting a harmless microbe in order to confer resistance to a dangerous one is called vaccination. Modern attempts to develop resistance to malaria[6], herpes[7], and other diseases often involve delivering antigens via a harmless vaccinia virus related to cowpox virus. 　　Many years passed before anyone learned how exposure to an infectious agent can confer resistance to a disease. A key step toward answering this question was taken more than a half-century later by the famous French scientist Louis Pasteur. Pasteur was studying fowl cholera[8], and he isolated a culture of bacteria from diseased chickens that would produce the disease if injected into	[1]感染 [2]水痘 [3]免疫学 [4]天花 [5]牛痘 [6]疟疾,瘴气 [7]疱疹 [8]鸡霍乱,家禽霍乱症

Text	Notes
healthy birds. Before departing on a two-week vacation, he accidentally left his bacterial culture out on a shelf. When he returned, he injected this old culture into healthy birds and found that it had been weakened; the injected birds became only slightly ill and then recovered. Surprisingly, however, those birds did not get sick when subsequently infected with fresh fowl cholera. They remained healthy even if given massive doses of active fowl cholera bacteria that did produce the disease in control chickens. Clearly, something about the bacteria could elicit immunity as long as the bacteria did not kill the animals first. We now know that molecules protruding from the surfaces of the bacterial cells evoked active immunity in the chickens. Key Concepts of Specific Immunity 　　An antigen is a molecule that provokes a specific immune response. Antigens are large, complex molecules such as proteins; they are generally foreign to the body, usually present of the surface of pathogens. A large antigen may have several parts, and each stimulate a different specific immune response. In this case, the different parts are known as antigenic determinant[9] sites, and each serves as a different antigen. Particular lymphocytes have receptor proteins on their surfaces that recognize an antigen and direct a specific immune response against either the antigen or the cell that carries the antigen. 　　Lymphocytes called B cells respond to antigens by producing proteins called antibodies. Antibody proteins are secreted into the blood and other body fluids and thus provide humoral immunity(The term humor here is used in its ancient sense, referring to a body fluid). Other lymphocytes called T cells do not secrete antibodies but instead directly attack the cells that carry the specific antigens. These cells are thus described as producing cell-mediated immunity. The specific immune responses protect the body in two ways.	[9]决定子,决定簇

Text	Notes
First, an individual can gain immunity by being exposed to a pathogen (disease-causing agent) and perhaps getting the disease. This is acquired immunity, such as the resistance to the chicken pox that you acquire after having the disease in childhood. Another term for this process is active immunity[10]. Second, an individual can gain immunity by obtaining the antibodies from another individual. This happened to you before you were born, with antibodies made by your mother being transferred to you across the placenta. Immunity gained in this way is called passive immunity[11].	[10] 自动免疫 [11] 被动免疫

Text A 参考译文

免疫反应

在生物体所面对的威胁中,有一部分是由细菌、病毒、真菌和其他外来的致病因子所引起的。所有的生物体都有非特异性免疫系统(或先天免疫系统),可以提供生物所需的部分保护功能。这种防御系统存在于从海绵动物到哺乳动物的整个动物界中。但是在脊椎动物中还有一种防御系统,称之为特异性免疫,这种免疫机制也被称为获得性免疫,或者简称为免疫反应。

综述

特异性免疫的一个特点是具有识别功能。当生物体识别出外来入侵物时,免疫反应就开始了,原因是外来细胞和自身体细胞具有不同的分子。能引起免疫反应的分子被称之为抗原。生物体通常不会对自身的抗原产生免疫反应,因为能够识别自身抗原的细胞已被去除或灭活。这种机制被称为自身耐受,它是定义免疫反应的一个关键特征。

第二个特征是特异性。尽管所有的免疫反应都比较相似,但是当机体受到不同的抗原入侵时,会产生与这种抗原相对应的免疫反应,例如,与脓毒性咽喉炎细菌引起的感染相比较,感冒病毒感染会引发不同种类细胞的反应。

第三个特点是具有记忆性。在抗原被从身体中清除以后,如果再遇到这种抗原,免疫记忆就会识别这种抗原,并能很快地将其清除出去。

抗原呈递

三个类群的白细胞与免疫反应的启动有关。尽管免疫反应可在身体任何一个具有这些细胞的部位发生,但最初只能在淋巴结和脾中发生。这些器官含有大量的抗原呈递细胞(APCs)、T 淋巴细胞(或 T 细胞)和 B 淋巴细胞(或 B 细胞)。

抗原呈递细胞包括巨噬细胞、树突细胞和 B 细胞。这些细胞遇到外来入侵物之后，把外来入侵物抗原呈递给一类称为辅助性 T 细胞（TH 细胞）的 T 细胞，抗原呈递细胞再把外来物体吞入，并把它带入细胞内。然后抗原呈递细胞把入侵物分解成抗原，再把这些抗原转移到细胞表面。受体是能够与抗原结合的细胞表面蛋白，每一个 TH 细胞都有不同的受体，这就使得不同的细胞只能识别特定的抗原。

抗原呈递细胞把抗原呈递给 TH 细胞，直到 TH 细胞受体与抗原相配套为止。两个细胞之间的接触会刺激 TH 细胞加速分裂，这个过程称之为克隆选择，因为只有识别了外来入侵物的 TH 细胞才会进行增殖。受到刺激的 TH 细胞也会产生一些称为细胞因子的化学信息素。所有的免疫细胞都会产生细胞因子，来调控免疫反应。

抗原清除

大量的辅助性 T 细胞激活了其他两种白细胞：细胞毒性 T 细胞（TC 细胞）和 B 细胞。与 TH 细胞相似，每一个 TC 细胞和 B 细胞都有与抗原相配套的受体，这就是为什么免疫系统能根据特异性识别几百万种抗原的原因。当具有配套受体的细胞遇到抗原时，就做好了激活的准备。它们收到从 TH 细胞和细胞因子发出的克隆选择所需的最终信号。克隆的 TC 细胞附着在它们识别出来的入侵物上，然后释放出多种化学物质来消灭外来细胞。由于必须要通过细胞与细胞的接触才能发生反应，因此这个反应就被称为细胞媒介免疫（或细胞免疫）。这对于消灭不正常的细胞非常有效，如癌细胞或病毒感染细胞。克隆的 B 细胞通过不同的方式来消灭入侵物。在 TH 细胞激活之后，B 细胞释放一种称之为抗体的蛋白。抗体通过体液流至全身各处，再与抗原结合，利用非特异性免疫机制消灭目标细胞。这种免疫反应我们称之为抗体媒介免疫（或体液免疫）。这种机制对于清除细菌、胞外病毒和其他存在于体液中的抗原特别有效。

免疫记忆

当机体首次遇到一种特异性抗体时，最初的免疫反应就发生了。启动反应需要几天时间，然后用 1 周到 2 周的时间来达到最大活性。如果以后机体再遇到同样的抗原，就会发生第二次免疫反应，这次只需要几个小时来启动免疫反应，而且可能在几天之内就会达到免疫反应的最大活性。这样外来入侵物在引起疾病之前就会被去除掉。这是因为在最初免疫反应发生时产生的一些克隆的 TC 细胞和 B 细胞发展成为了记忆细胞。如果同样的抗原出现，那么这些细胞就会立即被激活。前述的细胞间复杂的相互作用将不再需要。

实际上，当赋予一个生物体对某种疾病免疫力时，就会发生这样的反应。疫苗（利用弱毒或灭活病原体）首先引起最初的免疫反应（但不是疾病），然后产生的记忆细胞将会在遇到病原体时提供保护作用。

免疫系统失调

通过对免疫反应进行研究，科学家了解了免疫系统疾病，例如，超敏反应是指机体免疫系统对某种抗原所表现的异常增高的免疫应答，可损害健康组织。过多的抗体和 TC

细胞活动的结果可以是无害的(如对花粉、毒葛或霉菌过敏),也可能是致命的(如自身免疫系统疾病或对蜂毒和抗生素过敏)。与上述相反的另一个极端是免疫缺陷疾病,这种情况下机体不能有效地对入侵物产生免疫反应。HIV(人类免疫缺陷病毒)攻击 TH 细胞,其感染引起 AIDS(获得性免疫缺陷综合症)。有时一些个体在出生时就有免疫缺陷,但是这些异常通常是后天获得的(例如由于放射线辐射、化疗和 HIV 感染)。不管是什么原因,具有免疫缺陷的个体将很难抵抗感染。

Text B 参考译文

细胞反击:第二道防线

脊椎动物的体表防线是非常有效的,但是有时候也会被攻破,使得入侵物能进入到机体内部。在这种情况下,机体会运用一种非特异性的细胞和化学装置来进行防御。我们称之为第二道防线。这种装置都有一个共同的特性:不需要确定入侵物的类型,对任何的微生物感染都作出反应。

尽管非特异性免疫反应的细胞和化学物质会分散到全身各处,但是还有一个免疫中心来收集和分配免疫系统中的细胞,这个中心就是淋巴系统。淋巴系统包括一个由淋巴毛细管、淋巴管、淋巴结和淋巴器官组成的网络。尽管淋巴系统还具有与循环有关的其他功能,但它也能存储免疫反应所需的细胞和其他介质。这些细胞会被分配到全身各处以抗感染,这些细胞也存储在淋巴结上,在这里当体液通过时外来入侵物就会被清除。

杀灭入侵微生物的细胞

脊椎动物机体中最重要的非特异性免疫物质是称之为白细胞的血细胞,这种细胞通过血液循环到全身各处,然后攻击侵入体组织的微生物。白细胞有三种基本类型,每一种都可以杀灭不同的入侵微生物。

巨噬细胞是一种体积较大、形状不规则的细胞,通过噬菌作用把微生物吞入胞内,然后杀死微生物,与变形虫吞噬食物颗粒非常相似。在巨噬细胞内,由膜包被的液泡中含有溶酶体与细菌的融合体。这种融合激活了溶酶体中酶的活性,这些酶可通过释放大量的氧自由基来杀灭微生物。巨噬细胞也会吞噬病毒、细胞碎片和肺中的灰尘颗粒。巨噬细胞在细胞外液中不断地流动,巨噬细胞的噬菌作用是对肝、脾和骨髓结构一部分的特殊的噬菌细胞的作用的补充。对感染作出反应的过程中,血液中的单核细胞(未分化的白细胞)通过毛细血管进入结缔组织。在感染部位,单核细胞转化成巨噬细胞。

嗜中性粒细胞是一种与巨噬细胞相似的白细胞,通过噬菌作用吞入并杀灭微生物。另外,嗜中性粒细胞释放一些化学物质(有些与家用漂白剂相似),杀灭邻近的微生物和嗜中性粒细胞自身。

自然杀伤细胞不会直接攻击入侵的微生物,它们会杀灭机体中受病毒感染的细胞。它们不是通过噬菌作用,而是在目标细胞的细胞膜上打一个孔。自然杀伤细胞释放一种被称为孔形成蛋白的蛋白质,插入到目标细胞膜中,再形成一个孔。产生的孔会使得水分涌入目标细胞中,目标细胞膨胀然后破裂。自然杀伤细胞也会攻击癌细胞,一般是在癌细胞有机会发展成为可检测到的肿瘤之前。自然杀伤细胞灵敏的监测机制是身体最有效的防癌机制之一。

杀灭入侵微生物的蛋白质

一种称之为补助系统的化学防御机制有效地增强了脊椎动物的细胞防御能力。这个系统包括20来种在血浆中自由循环的蛋白质。当它们遇到细菌或真菌细胞壁时，这些蛋白质集合成膜攻击复合体，插入到外来细胞的质膜中，形成一个与杀伤细胞作用相类似的孔，水分再由这个孔进入到细胞内，引起细胞膨胀和破裂。抗体结合到入侵的微生物中也会激发起补体蛋白的凝集。

补体系统的的蛋白质能增强身体其他防御系统的作用。有些补体通过组胺的释放增强炎症反应；其他补体会吸引巨噬细胞赶赴受感染位点；还有一些能包在入侵的微生物表面，使其表面变粗糙，从而有利于噬菌细胞黏附在它们表面。

另有一些与身体防御有关的蛋白质是干扰素。有三种主要的干扰素：α、β和γ干扰素。机体中几乎所有的细胞都产生α、β干扰素。这些多肽起着信息素的作用，能保护邻近受感染细胞的正常细胞免受感染。尽管病毒仍然能够穿透邻近细胞，但α、β干扰素能够防止病毒在这些细胞中的复制和病毒蛋白质的装配。γ干扰素只由特殊的淋巴细胞和自然杀伤细胞生成。这些细胞中γ干扰素的分泌是免疫防御中抗感染和抗癌机制的组成部分。

Unit 10

Text A

Stem Cells

Stem cells are cells that grow and divide without limitation, and given the proper signals, can become any other type of cell.

Some stem cells are embryonic in origin. As a human embryo grows, the early cells start dividing and forming different, specialized cells such as heart cells, bone cells, and muscle cells. Once formed, specialized nonstem cells can only divide to produce replicas of themselves. They cannot backtrack and become a different type of cell.

Embryonic stem cells retain the ability to become virtually any cell type. If the cells are harvested from an early embryo (about 5-7 days after conception) and nudged in a particular direction in the laboratory, they can be directed to become a particular tissue or organ.

Tissues and organs grown from stem cells in the laboratory may some day be used to replace organs damaged in accidents or organs that are gradually failing due to degenerative diseases. Degenerative diseases start with the slow breakdown of an organ and progress to organ failure. Additionally, when one organ is not working properly, other organs are also affected. Degenerative diseases include stroke, diabetes, liver and lung diseases, heart disease, and Alzheimer's disease.

Stem cells could provide healthy tissue to replace those damaged by spinal cord injury or burns. New heart muscle could be produced to replace that damaged during a heart attack. A diabetic could have a new pancreas, and people suffering from osteoarthritis could have replacement cartilage to cushion their joints. Thousands of people waiting for organ transplants might be saved if new organs were grown in the lab.

One problem with stem-cell research is that the embryos are destroyed when the stem cells are removed. And many people object to the destruction of early embryos. Currently, the federal government will fund research using leftover embryos from fertility treatments, but will not support research using embryos created solely for research purposes. This ban only applies to federally funded research projects, which means that in the United States, research on embryos can only be performed by genetic engineers who obtain grants from nongovernmental sources unless they have access to the limited numbers of embryos created during fertility treatments.

In-vitro (Latin, meaning "in glass") fertilization procedures often result in the production of excess embryos because a large number of egg cells are harvested from a woman who wishes to become pregnant. These egg cells are then mixed with her partner's sperm in a petri dish, resulting in the production of many fertilized eggs that grow into embryos. A few of the embryos are then implanted into the woman's uterus. The remaining embryos are stored so that more attempts can be made if pregnancy does not result or if the couple desires more children. When the couple achieves the desired number of pregnancies, the remaining embryos can, with the couple's consent, be used for stem-cell research.

A solution to the ethical dilemma presented by the use of embryonic stem cells seems to be on the horizon. Scientists have recently discovered that many adult tissues also contain stem cells. Recent studies published in peer-reviewed literature suggest that most adult tissues have stem cells, that these cells can be driven to become other cell types, and that they can be grown indefinitely in the laboratory. Based on success in animal models, there is even evidence that adult stem cells will help cure diseases. In fact, scientists have used stem cells from adult tissues to repair damage in animals due to heart attack, stroke, diabetes, and spinal cord injury.

New Words

limitation	[ˌlimiˈteiʃən]	n.	限制
signal	[ˈsignəl]	n.	信号
replica	[ˈreplikə]	n.	复制品
backtrack	[ˈbæktræk]	vi.	返回,退回
harvest	[ˈhɑːvist]	v.	收获,获得,收集
		n.	收获,收成,结果,成果
nudge	[nʌdʒ]	vt.	轻触;轻推
stroke	[strəuk]	n.	中风
diabetes	[daiəˈbiːtiːz]	n.	糖尿病
pancreas	[ˈpænkriəs]	n.	胰腺
osteoarthritis	[ɔstiːˌəuɑːˈθraitis]	n.	骨关节炎
cushion	[ˈkuʃən]	vt.	放在
pregnant	[ˈpregnənt]	adj.	怀孕的,怀胎的
implant	[ˈimplɑːnt]	vt.	植入

uterus	[ˈjuːtərəs]	n. 子宫
pregnancy	[ˈpregnənsi]	n. 怀孕
consent	[kənˈsent]	vi. 同意，赞成，答应
		n. 同意，赞成，允诺
ethical	[ˈeθikəl]	adj. 伦理的；道德的
dilemma	[daiˈlemə]	n. 进退两难之境；困境

Phrases

stem cell	干细胞
heart cell	心肌细胞
degenerative disease	退行性疾病
Alzheimer's disease	阿尔茨海默病
organ transplant	器官移植
object to	反对
have access to	有机会得到；能接近；能使用
in-vitro fertilization	体外受精
egg cell	卵细胞
petri dish	培养皿
fertilized egg	受精卵

Notes

[1] Stem cells are cells that grow and divide without limitation, and given the proper signals, can become any other type of cell.

　　本句中，that grow and divide without limitation 是一个定语从句，修饰和限定 cells。given the proper signals 是一个过去分词短语，作条件状语，修饰谓语 can become。can become any other type of cell 前面省略了主语 Stem cells。

[2] If the cells are harvested from an early embryo (about 5-7 days after conception) and nudged in a particular direction in the laboratory, they can be directed to become a particular tissue or organ.

　　本句中，If the cells are harvested from an early embryo (about 5-7 days after conception) 是由 if 引导的条件状语从句，nudged in a particular direction in the laboratory 也是一个 if 引导的条件状语从句，它前面省略了 If the cells are, and 是个连词，起连接作用。

[3] Tissues and organs grown from stem cells in the laboratory may some day be used

to replace organs damaged in accidents or organs that are gradually failing due to degenerative diseases.

本句中,grown from stem cells in the laboratory 是一个过去分词短语,作定语,修饰和限定主语 Tissues and organs。damaged in accidents 也是一个过去分词短语,作定语,修饰和限定它前面的 organs,that are gradually failing due to degenerative diseases 是一个定语从句,修饰和限定它前面的 organs,其中 due to degenerative diseases 作原因状语。

[4] This ban only applies to federally funded research projects, which means that in the United States, research on embryos can only be performed by genetic engineers who obtain grants from nongovernmental sources unless they have access to the limited numbers of embryos created during fertility treatments.

本句中,which means that in the United States, research on embryos can only be performed by genetic engineers who obtain grants from nongovernmental sources unless they have access to the limited numbers of embryos created during fertility treatments 是一个非限定性定语从句,对主句作进一步补充说明。在该从句中 that 引导了一个宾语从句,who obtain grants from nongovernmental sources 是一个定语从句,修饰和限定 genetic engineers。unless they have access to the limited numbers of embryos created during fertility treatments 是一个条件状语从句。

Exercises

【EX. 1】 根据课文内容,回答以下问题

1) What are stem cells?

2) What do the early cells do as a human embryo grows?

3) What can be done to cells if they are harvested from an early embryo (about 5-7 days after conception) and nudged in a particular direction in the laboratory?

4) What do degenerative diseases do?

5) What do degenerative diseases include?

6) What could stem cells provide?

7) What is the one problem with stem-cell research?

8) What kind of research will the federal government fund? What kind don't they support?

9) What do in vitro (Latin, meaning "in glass") fertilization procedures often result in?

10) What do recent studies published in peer-reviewed literature suggest?

【EX. 2】 根据下面的英文解释,写出相应的英文词汇

英　文　解　释	词　汇
A copy or reproduction, especially one on a scale smaller than the original.	
Carrying developing offspring within the body.	
A hollow muscular organ located in the pelvic cavity of female mammals in which the fertilized egg implants and develops.	
To insert or embed surgically.	
A sudden loss of brain function caused by a blockage or rupture of a blood vessel to the brain, characterized by loss of muscular control, diminution or loss of sensation or consciousness, dizziness, slurred speech, or other symptoms that vary with the extent and severity of the damage to the brain.	
Any of several metabolic disorders marked by excessive discharge of urine and persistent thirst.	
A form of arthritis, occurring mainly in older persons, that is characterized by chronic degeneration of the cartilage of the joints.	
A long, irregularly shaped gland in vertebrates, lying behind the stomach, that secretes pancreatic juice into the duodenum and insulin, glucagon, and somatostatin into the bloodstream.	
A situation that requires a choice between options that are or seem equally unfavorable or mutually exclusive.	
Agreement as to opinion or a course of action.	

【EX. 3】 把下列句子翻译为中文

1) Recent experiments have demonstrated the possibility of cloning differentiated

mammalian tissue, opening the door for the first time to practical transgenic cloning of farm animals.

2) Transplanted stem cells may allow us to replace damaged or lost tissue, offering cures for many disorders that cannot now be treated. Current work focuses on tissue-specific stem cells, which do not present the ethical problems that embryonic stem cells do.

3) Recent experiments open the way for cloning of genetically altered animals and suggest that human cloning is feasible.

4) Crop foods may be genetically modified to increase their yield, shelf life, and nutritive content.

5) There is concern that GM foods may negatively affect the environment or people who consume them.

6) Cloning animals with desirable agricultural traits has occurred. It may someday be possible to clone humans, but it is unclear if these humans would be healthy.

7) Human cloning occurs commonly in nature via the spontaneous production of identical twins.

8) Cloning offspring from adults with desirable traits has been successfully performed on cattle, goats, mice, cats, pigs, rabbits, and sheep.

9) Since the isolation of embryonic stem cells in 1998, labs all over the world have been exploring the possibility of using stem cells to restore damaged or lost tissue.

10) The exciting promise of these embryonic stem cells is that, because they can develop into any tissue, they may give us the ability to restore damaged heart or spine tissue.

【EX. 4】 把下列短文翻译为中文

Wilmut's successful cloning of fully differentiated sheep cells is a milestone event in gene technology. Even though his procedure proved inefficient (only one of 277 trials succeeded), it established the point beyond all doubt that cloning of adult animal cells can be done. In the following four years researchers succeeded in greatly improving the efficiency of cloning. Seizing upon the key idea in Wilmut's experiment, to clone a resting-stage cell, they have returned to the nuclear transplant procedure pioneered by Briggs and King. It works well. Many different mammals have been successfully cloned including mice, pigs, and cattle.

Transgenic cloning can be expected to have a major impact on medicine as well as agriculture. Animals with human genes can be used to produce rare hormones. For example, sheep that have recently been genetically engineered to secrete a protein called alpha-1 antitrypsin (helpful in relieving the symptoms of cystic fibrosis) into their milk may be cloned, greatly cheapening the production of this expensive drug.

Text B

Gene Therapy

Gene therapy is an experimental disease treatment in which a gene is delivered to cells in the body. The protein made by the new gene compensates for the absence of normal proteins or interacts with some abnormal protein already in the cell to interrupt its function. Gene therapy is not yet a routine treatment for any disease, but it may become so as researchers solve the many technical problems it presents.

Humans are prey to numerous diseases due to single-gene defects, such as adenosine deaminase deficiency (defective enzyme), cystic fibrosis (defective ion channel), and Duchenne muscular dystrophy (defective muscle protein). Replacement of the defective gene is conceptually simple, but practically very difficult. Effective gene therapy requires delivering the gene to each cell in which it acts, integrating the gene with the thousands of others on the chromosomes and regulating the expression of the gene.

Gene delivery is a major hurdle. Viruses are the most commonly used vehicles, or vectors, since they have been designed by evolution to deliver their own genes to our cells. Adenovirus (a type of cold virus) has been the most commonly used vector, since it can carry a very large gene and will infect most cell types. However, the immune system is designed to prevent this type of infection, and immune rejection has so far thwarted most gene therapy efforts. While most patients have not been harmed by this problem, one gene therapy patient has died from immune response to the adenovirus. Modifications of the virus, using fewer immunogenic viruses (such as adeno-associated virus, herpes virus, or retrovirus), immune-suppressive drugs, and nonviral delivery systems are all possible solutions. Curiously, the brain does not mount a strong immune response, and as such, represents a promising site for gene delivery in neurological diseases.

Getting the gene to enough target cells is also a significant challenge. Adenosine deaminase deficiency affects white blood cells and causes severe combined immune deficiency ("bubble boy" disease). This disease can be treated by removing white blood cells, inserting the adenosine deaminase gene into them, and returning the cells to the bone marrow. Cystic fibrosis presents a much bigger challenge, since it affects the airways and pancreas. Inhalation of the vector may treat the lungs, but the pancreas is more difficult to reach without injecting vector into the bloodstream. Duchenne muscular dystrophy is an even bigger challenge, since it affects all muscles, and muscles make up 45 percent of the body. The only realistic treatment option in this case is systemic delivery, which poses the added challenge of preventing delivery to nonmuscle tissue.

Once inside the target tissue, genes usually become active whether or not they are integrated into the host chromosome. However, long-term expression requires that the gene join the host chromosome. Directing the gene to do so and integrating it in a way that doesn't disrupt other genes is still a significant challenge. Regulating its expression so that enough of the protein (but not too much) is made is also a problem. Currently, most virally delivered genes do not integrate successfully and stop making protein after several weeks to months.

While correction of gene defects was the original inspiration for gene therapy research, treatment of other diseases is now being explored. Cancers are an appealing target, and several strategies are possible. Currently the most promising is delivering a so-called "suicide gene", whose protein product renders a tumor more sensitive to cell-killing drugs, allowing lower doses of chemotherapy to be effective. This works well for solid tumors, which can be injected with the gene. Delivery to more diffuse locations is still problematic. Further research on cellular properties of cancer cells may broaden the reach of this and similar cancer-targeting strategies.

New Words

therapy	['θerəpi]	n.	治疗
compensate	['kɔmpənseit]	v.	偿还,补偿
absence	['æbsəns]	n.	缺乏,没有
routine	[ruː'tiːn]	adj.	例行的;常规的
		n.	常规;惯例
hurdle	['həːdl]	n.	障碍
vehicle	['viːikl]	n.	工具;载体
vector	['vektə]	n.	带菌体,载体
adenovirus	[ˌædinəu'vaiərəs]	n.	腺病毒;呼吸系统病毒
thwart	[θwɔːt]	vt.	阻挠;妨碍;使挫折

immunogenic	[ˌimjuːnəˈdʒenik]	adj. 产生免疫性的,免疫原的
unviral	[ʌnˈvairəl]	adj. 非滤过性毒菌的,非滤过性毒菌引起的
neurological	[ˌnjuərəˈlɔdʒikəl]	adj. 神经学上的,神经病学的
airway	[ˈɛəwei]	n. 气管,空气道
inhalation	[ˌinhəˈleiʃən]	n. 吸入
disrupt	[disˈrʌpt]	v. 干扰,破坏
render	[ˈrendə]	vt. 给;使,致使
chemotherapy	[ˌkeməˈθerəpi]	n. 化学疗法

Phrases

adenosine deaminase deficiency	腺苷脱氨(基)酶缺乏症
cystic fibrosis	囊性纤维化,囊性纤维变性,胞囊纤维化
Duchenne muscular dystrophy	杜兴肌营养不良症
gene delivery	基因导入
immune rejection	免疫排斥

Exercises

【EX. 5】 根据课文内容,回答以下问题

1) What is gene therapy?

2) Is gene therapy a routine treatment for any disease?

3) What does effective gene therapy require?

4) What is the reason given in the text to explain that viruses are the most commonly used vehicles to gene delivery?

5) Why has Adenovirus been the most commonly used vector?

6) What does adenosine deaminase deficiency do?

7) How can immune deficiency ("bubble boy" disease) be treated?

8) Why does cystic fibrosis present a much bigger challenge?

9) Why is Duchenne muscular dystrophy an even bigger challenge? And what is the only realistic treatment option in this case?

10) What does long-term expression require?

Reading Material

Text	Notes
Should We Label Genetically Modified Foods 　　While there seems little tangible risk in the genetic modification of crops, public assurance that these risks are being carefully assessed[1] is important. Few issues manage to raise the temperature of discussions about plant genetic engineering[2] more than labeling of genetically modified[3] (GM) crops. Agricultural producers have argued that there are no demonstrable[4] risks, so that a GM label can only have the function of scaring off wary consumers. Consumer advocates respond that consumers have every right to make that decision, and to the information necessary to make it. 　　In considering this matter, it is important to separate two quite different issues, the need for a label, and the right of the public to have one. Every serious scientific investigation of the risks of GM foods has concluded that they are safe—indeed, in the case of soybeans and many other crops modified to improve cultivation[5], the foods themselves are not altered in any detectable way, and no nutritional test could distinguish them from "organic" varieties. So there seems to be little if any health need for a GM label for genetically engineered foods. 　　The right of the public to know what they are eating is a very different issue. There is widespread fear of genetic manipulation in Europe, because it is unfamiliar. People there don't trust their regulatory agencies[6] as we do here, because their agencies have a poor track record of protecting them. When they look at genetically modified foods, they are haunted[7] by past experiences of regulatory	[1]估定,评定 [2]基因工程 [3]转基因的 [4]可论证的,可表明的;显而易见的 [5]耕种,耕作 [6]制定规章的机构 [7]令人烦恼的,困惑的

Text	Notes
ineptitude[8]. In England they remember British regulators' failure to protect consumers from meat infected with mad cow disease[9].	[8]不称职 [9]疯牛病
It does no good whatsoever to tell a fearful European that there is no evidence to warrant fear, no trace of data supporting danger from GM crops. A European consumer will simply respond that the harm is not yet evident, that we don't know enough to see the danger lurking[10] around the corner. "Slow down", the European consumers say. "Give research a chance to look around all the corners. Let's be sure." No one can argue against caution, but it is difficult to imagine what else researchers can look into—safety has been explored very thoroughly. The fear remains, though, for the simple reason that no amount of information can remove it. Like a child scared of a monster[11] under the bed, looking under the bed again doesn't help—the monster still might be there next time. And that means we are going to have to have GM labels, for people have every right to be informed about something they fear.	[10]潜伏,埋伏 [11]怪物,妖怪
What should these labels be like? A label that only says "GM FOOD" simply acts as a brand—like a POISON label, it shouts a warning to the public of lurking danger. Why not instead have a GM label that provides information to the consumer, that tells the consumer what regulators know about that product? (For Bt corn): The production of this food was made more efficient by the addition of genes that made plants resistant[12] to pests so that less pesticides were required to grow the crop. (For Roundup-ready soybeans): Genes have been added to this crop to render it resistant to herbicides—this reduces soil erosion[13] by lessening the need for weed removing cultivation. (For high beta-carotene rice): Genes have been added to this food to enhance its beta-carotene content and so combat vitamin A deficiency. GM food labels that in each instance[14] actually tell consumers what has been done to the gene-modified crop would go a long way toward hastening[15] public acceptance of gene technology in the kitchen.	[12]抵抗的,有抵抗力的 [13]侵蚀土壤 [14]情况 [15]催促,加速,加快

Text A 参考译文

干 细 胞

  干细胞是一群具有无限生长和分裂能力的细胞,如果有正确的细胞信号,干细胞能发育成任何类型的细胞。

  有些干细胞源自胚胎。随着人类胚胎的发育,早期的细胞开始分化并形成不同类型的细胞,如心肌细胞、骨细胞和肌细胞。一旦分化以后,这些非干细胞就只能进行复制,不会再恢复为干细胞而成为其他类型的细胞。

  胚胎干细胞具有分化成其他任何类型细胞的能力,如果从一个早期的胚胎中收集这种细胞(大约在受孕后 5 天),然后在实验中对其分化方向进行控制,它们就能直接发育成为某一特定的组织和器官。

  将来总有一天,在实验室中利用干细胞培养出来的组织和器官可能可以替代已损坏的组织和器官——这些组织和器官可能在一场事故中受到损坏,也可能由于退行性疾病造成器官逐渐衰竭。此外,如果一个器官功能不正常,其他的器官也会受到影响。退行性疾病包括中风、糖尿病、肝和肺的疾病、心脏病和阿尔茨海默病。

  干细胞能提供健康的组织来替代脊髓受损组织和烧伤的组织。新生成的心肌也可替代心脏受损组织。糖尿病患者可能会有一个新的胰腺,那些正受到骨关节炎病痛折磨的人也可能会有可替代的软骨重新植入他们的关节中。如果新的器官能在实验室中生产出来,将会解救许多正在等待进行器官移植的人。

  干细胞研究中出现的一个问题是当干细胞被移出时,胚胎会受到破坏。而许多人反对破坏早期胚胎。目前,联邦政府将资助那些采用受精后剩余胚胎的研究项目,如果在一项研究中,所制备的胚胎单单只是为了研究,那么就不会获得资助。这项规定只适用于联邦基金,这就意味着在美国,进行胚胎研究的科学家只能从非政府机构获得基金资助,除非他们有机会得到数量有限的受精后剩余的胚胎。

  体外(拉丁语,意思是"在试管中")受精步骤通常产生多余的胚胎,因为将有大量想怀孕的妇女捐献卵细胞。将这些妇女卵细胞与她们伴侣的精子在培养皿中混合,结果产生大量的受精卵,这些受精卵最后会发育成胚胎。一些胚胎会被重新植回到子宫里,其余的胚胎将会被保存起来,这样,如果没有怀孕成功或一对夫妻想要更多的孩子,则再进行一次胚胎植入。而当一对夫妻达到目的以后,其余的胚胎在得到当事人的允许之后,将被用于干细胞研究。

  似乎已经找到使用胚胎干细胞出现的伦理问题的解决方案。最近,科学家发现成人的许多组织中也含有干细胞。最近发布的同行评审研究成果表明,大多数成人组织中具有干细胞,这些干细胞同样也能分化成其他类型细胞,而且在实验中这些干细胞能无限制地生长。在成功的动物模型中,甚至有证据表明成人干细胞将有助于治疗疾病。实际上,科学家已经利用从成人组织中分离到的干细胞对有心脏病、中风、糖尿病和脊髓损伤的动物机体进行修复。

Text B 参考译文

基 因 治 疗

　　基因治疗是一种试验性的疾病治疗方法,需要把基因导入体细胞,然后由新基因产生的蛋白质就会修补正常蛋白质的缺失,或通过与细胞中异常蛋白质的相互作用而影响异常蛋白质的功能。对于任何一种疾病来说,基因治疗还不是一种常规的治疗方法,但是当研究人员解决了它所存在的许多技术问题之后,它将会成为一种常规的方法。

　　由于单个基因的缺失,许多人正受到疾病的折磨,比如腺苷脱氨酶的缺乏症(酶的欠缺)、胞囊纤维化(离子通道的缺失)和肌肉的杜兴肌营养不良症(缺乏肌肉蛋白质)。在理论上替换有缺陷的基因是简单的,但是实际操作却非常困难。有效的基因治疗需要将基因导入到每一个起作用的细胞中,使导入基因与染色体上的其他基因整合在一起,然后再调控基因的表达。

　　基因导入是一个主要的难题,病毒是一个常用的导入工具或载体,因为它们经过进化已经具有了把它们自身的基因导入到细胞中的能力。腺病毒(一种感冒病毒)是最常用的一种载体,因为它携带有一个大的基因,且能感染大多数类型的细胞。但是,免疫系统会阻止这种感染,因此,迄今为止,免疫排斥会对大多数基因治疗效果产生负面影响。多数病人未受到这个问题的伤害,但有一位基因治疗的病人死于对腺病毒的免疫反应。对病毒进行改造、使用免疫原较少的病毒(如腺伴随病毒、疱疹病毒或逆转录酶病毒)、使用免疫抑制药物和非滤过性毒菌导入系统都是解决问题的可能方法。奇怪的是,大脑不会产生很强的免疫反应,正因为如此,大脑为神经系统疾病的治疗提供了一个有希望的基因导入位点。

　　把基因导入到足够多的靶细胞中去也是一个很具有挑战性的工作。腺苷脱氨酶缺乏症影响了白细胞,并引起严重的免疫缺乏症("泡泡男孩"病)。这种病的治疗方法是移出白细胞,把腺苷脱氨酶基因植入到白细胞中,再让它们回到骨髓中去。胞囊纤维化则更具有挑战性,因为这种病会影响气管和胰腺。肺可以用吸入载体的方法,但是胰腺则比较困难,它需要把载体注射到血液中。杜兴肌营养不良症操作更为困难,因为它影响所有的肌肉,而肌肉占身体的 45%。这种情况下最可行的治疗方法是系统导入,但是这也增加了一个难题,即如何防止导入到非肌肉组织中去。

　　一旦进入到目标组织中,基因一般都会被激活,而不管它们是否已被整合到主染色体中。但是,如果基因要持续表达,必须要让基因整合到主染色体中。检测基因是否已被整合以及整合时会不会影响其他的基因,仍然是个大的挑战。调控基因的表达以产生足够多的蛋白质(但不是太多)也是一个问题。当前,多数以病毒方式导入的基因都没有整合成功,几周或几个月之后,就会停止产生蛋白质。

　　纠正基因缺失为基因治疗研究提供了原动力,同时对其他疾病的治疗也正在研究。癌症是一个吸引人的研究目标,几种方法对癌症治疗都是可能的。目前最有希望的方法是导入一种称之为"自杀基因"的基因,它产生的蛋白质增加肿瘤对杀伤细胞药物的敏感度,使得化学疗法中药品的使用量更少、更有效。这对于固体肿瘤具有很好的作用,因为可以使用基因注射的方法。导入到比较容易扩散的部位依然是个问题。对癌细胞特性的进一步研究将会拓宽解决这个问题和类似的定位癌细胞策略的视野。

练习答案

Unit 1

【EX. 1】 根据课文内容,回答以下问题

1) Biology literally means "the study of life".
2) It covers the minute workings of chemical machines inside our cells, as well as the concepts of ecosystems and global climate change.
3) Biologists study intimate details of the human brain, the composition of our genes, and even the functioning of our reproductive system.
4) Deoxyribonucleic acid(DNA)is a chemical that had (then) recently been deduced to be the physical carrier of inheritance.
5) The cell theory states that all organisms are composed of one or more cells, and that those cells have arisen from pre-existing cells.
6) James Watson and Francis Crick developed the model for deoxyribonucleic acid, it was in 1953.
7) Homeostasis is the maintenance of a dynamic range of conditions within which the organism can function.
8) The major components of homeostasis are temperature, pH and energy.
9) Theromodynamics is a field of study that covers the laws governing energy transfers, and thus the basis for life on earth.
10) The two major laws mentioned in the passage are the conservation of matter and energy, and entropy.

【EX. 2】 根据下面的英文解释,写出相应的英文词汇(使用本单元所学的单词、词组或缩略语)

biology; cell; predisposition; informercial; replication; heredity; carrier; homeostasis; ecosystem; deduce

【EX. 3】 把下列句子翻译为中文

1) 生物学的核心理论是:生物多样性是长期进化的结果。

2) 科学是对客观信息进行观察和对这种信息进行研究以建立理论的一种方法。
3) 所有的生物都有一些主要的共同特征：有序、刺激感受性、生长、发育及繁殖、自我调节和动态平衡。
4) 生物具有不同等级水平的高度组织性，不管是一个简单的细胞还是复杂的组织都是如此。
5) 科学家和非专业人员都被生物学所吸引，因为它试图解释生命是如何起源的。
6) 几乎所有情况下，两种多细胞生物亲缘关系越近，它们在各方面的解剖结构越相似。
7) 在生物学上着重强调的是用生物体生产产品的生物技术。
8) 同时，由于生物技术的运用需要科学家操纵进化的历程，将来，科学家们将面对严肃的伦理观念的挑战。
9) 事实上，从最小的细菌到蓝鲸和巨大的美洲杉，每一个生物都用同样的遗传密码来制造蛋白质。
10) 科学家用各种技术和理论上的手段来研究生物。

【EX. 4】 把下列短文翻译为中文

　　生物学是一门有趣而重要的学科，因为它显著地影响着我们的日常生活和未来。许多生物学家正在研究对我们生活有重大影响的问题，如快速膨胀的世界人口以及癌症和艾滋病等疾病。生物学家获得的知识从某种程度上来说是我们维持世界资源、防治疾病、改善我们以及子孙后代生活质量等能力的基础。

　　生物学是最成功地解释我们的世界是什么的自然科学之一。为了解生物学，你必须先了解科学的本质。科学家使用的基本工具是思想。为了解科学的本质，不妨注意一下科学家的思考方式。他们以两种方式思考：演绎和归纳。

【EX. 5】 根据课文内容，回答以下问题

1) Homeostasis is the maintenance of a constant yet also dynamic internal environment in terms of temperature, pH, water concentrations, etc.
2) Most living things use the chemical DNA (deoxyribonucleic acid) as the physical carrier of inheritance and the genetic information.
3) Multicellular organisms pass through a more complicated process of differentiation and organogenesis because they have so many more cells to develop.
4) There are four classes of macromolecules in living things. They are polysaccharides, triglycerides, polypeptides and nucleic acids.
5) Sugars are structurally the simplest carbohydrates. They are the structural unit which makes up the other types of carbohydrates.
6) A catalyst is a chemical that promotes a chemical reaction but is not changed by it.
7) The building block of any protein is the amino acid, which has an amino end (NH_2) and a carboxyl end (COOH).

8) Nucleic acids are polymers composed of monomer units known as nucleotides.

9) The main functions of nucleotides are information storage (DNA), protein synthesis (RNA), and energy transfers (ATP and NAD).

10) Nucleotides consist of a sugar, a nitrogenous base, and a phosphate.

Unit 2

【EX. 1】 根据课文内容,回答以下问题

1) When the chromosome condensation initiated in G2 phase reaches the point at which individual condensed chromosomes first become visible with the light microscope, the first stage of mitosis, prophase, has begun.

2) Ribosomal RNA synthesis ceases when the portion of the chromosome bearing the rRNA genes is condensed.

3) Each chromosome possesses two kinetochores.

4) Because microtubules extending from the two poles attach to opposite sides of the centromere, they attach one sister chromatid to one pole and the other sister chromatid to the other pole.

5) Yes, it is. If the two sides of a centromere are attached to the same pole, for example, it will lead to a failure of the sister chromatids to separate, so that they end up in the same daughter cell.

6) No. The metaphase plate is not an actual structure. It is an indication of the future axis of cell division.

7) Of all the stages of mitosis, anaphase is the shortest and the most beautiful to watch.

8) Because another group of microtubules attach the chromosomes to the poles, the chromosomes move apart.

9) When the sister chromatids separate in anaphase, the accurate partitioning of the replicated genome is complete.

10) One of the early group of genes expressed are the rRNA genes.

【EX. 2】 根据下面的英文解释,写出相应的英文词汇(使用本单元所学的单词、词组或缩略语)

mitosis; chromosome; centriole; anaphase; microtubule; prophase; initiate; centromere; ribosomal; elongated

【EX. 3】 把下列句子翻译为中文

1) 所有的细胞都源自先前已存在的细胞,并有相同的特定的过程、分子类型和结构。

2) 第一个细胞可能来自气泡中大分子的聚合。
3) 为维持与环境进行适当的交换,细胞的表面积相对于其体积而言必须比较大。
4) 所有的原核细胞都有一个质膜、一个含 DNA 的核区以及一个含核糖体、水分、溶解蛋白质和小分子的细胞质。
5) 与原核细胞一样,真核细胞有一个细胞膜、细胞质和核糖体。不同的是真核细胞较大,并含有许多被膜包裹的细胞器。
6) 细胞核含有细胞中大部分 DNA,它与蛋白质结合形成染色体。
7) 多细胞生物通常由许多小细胞而不是一些大细胞组成,因为小细胞作用效率更高。它们的相对表面积较大,细胞中心与环境的交换速率更高。
8) 细菌是缺乏内部组织的小细胞。它们被由短多肽与碳水化合物交叉连接组成的细胞壁包围,有些细菌受旋状的鞭毛推动。
9) 多细胞二倍体或多细胞单倍体个体都是从有丝分裂开始时的单细胞发育而来的。
10) 核糖体是细胞质中蛋白质合成的地方。

【EX.4】 把下列短文翻译为中文

具有额外整套染色体的有机体有时可通过人工育种或自然的例外情况产生。有时候,三倍体(3n)、四倍体(4n)和更高的多倍体核可能形成。这些倍数分别代表存在的整套染色体数量的增加幅度。如果一个细胞核有一套或多套染色体,它本身异常的高倍体性并不会阻止有丝分裂的进行。在有丝分裂过程中,每一条染色体行为都不依赖于其他染色体。相反,在减数分裂过程中同源染色体必须联合后再分裂。如果一条染色体没有姐妹染色体,在细胞分裂的后期 I 不能将那条染色体拉到两极。双倍体细胞核可以进行正常的减数分裂;而单倍体则不能。同样,四倍体核的每一种染色体是偶数的,所以每一条染色体能与它的姐妹染色体配对。但是三倍体核不能进行普通的减数分裂,因为三分之一染色体没有它相应的姐妹染色体。

【EX.5】 根据课文内容,回答以下问题

1) Cells are the basic units of living organisms, with the exception of viruses whose structure and function are different from cells.
2) All cells are divided into two types: prokaryotic cells and eukaryotic cells. The prokaryotic cell does not have a nucleus. The eukaryotic cell contains a nucleus.
3) The prokaryotic cell does not have a nucleus. The eukaryotic cell contains a nucleus.
4) Eukaryotes are the organisms made up of eukaryotic cells.
5) Eukaryotes include protista, fungi, animals and plants. Prokaryotes include archaebacteria and eubacteria.
6) Methanogens live in anaerobic environment such as swamps.
7) Extreme halophiles live in very high concentrations of salt (NaCl), e.g., the Dead Sea and the Great Salt Lake.

8) Extreme thermophiles live in hot, sulfur rich and low pH environment, such as hot springs, geysers and fumaroles in the Yellowstone National Park.

9) The cell membrane functions as a semi-permeable barrier, allowing a very few molecules across it while fencing the majority of organically produced chemicals inside the cell.

10) No, not all living things have cell walls, most notably animals and many of the more animal-like protistans have not cell walls. Bacteria have cell walls containing peptidoglycan. Cellulose is the most common chemical in the plant primary cell wall.

Unit 3

【EX. 1】 根据课文内容,回答以下问题

1) Genes are segments of the DNA that code for specific proteins. These proteins are responsible for the expression of the phenotype.

2) Codominant alleles occur when the heterozygotes express both homozygous phenotypes.

3) Blood Type A people manufacture only anti-B antibodies and type B people make only anti-A antibodies.

4) Incomplete dominance is a condition when neither allele is dominant over the other.

5) Multiple alleles result from different mutations of the same gene.

6) The only possible genotype for a type O person is OO. Type A people have either AA or AO genotypes. Type B people have either BB or BO genotypes. Type AB people have only the AB (heterozygous) genotype.

7) Novel phenotypes often result from the interactions of two genes.

8) Epistasis is the term appl produced only by the rrpp genotype.

9) Phenotypes are always affected by their environment.

10) Expression of phenotype is a result of interaction between genes and environment.

【EX. 2】 根据下面的英文解释,写出相应的英文词汇

phenotype; heterozygote; homozygous; mutation; recessive; trait; gene; antigen; glycoprotein; dihybrid

【EX. 3】 把下列句子翻译为中文

1) 父母亲把他们等位基因中的独特的一套遗传给他们每一个不同的双胞胎子女。

2) 带有特殊基因的个体的表现型依赖于它携带哪个等位基因(它的基因型)以及这些等位基因是显性的、隐性的还是等显性的。

3) 基因在决定数量性状表现型中的作用是通过计算这种性状的遗传率来估计的。

4) 虽然孟德尔的研究工作是严谨的,并有很好的文档记载,他发表于19世纪60年代的发现直到数十年以后才引起人们注意。
5) 显性有时候是不完全的,这是因为在杂合子有机体中的两个等位基因都可以在表现型中表达。
6) 基因以多肽(蛋白质)形式在表现型中表达。
7) 人类中有一些遗传疾病已经被发现是由于缺乏一些酶造成的。这些发现支持了一个基因一个多肽学说。
8) DNA突变常常以异常的蛋白质表现。然而,表现型的变化并不容易被发现。有些突变仅仅在一定条件下才表现。
9) 染色体突变(缺失、重复、倒置或易位)发生在一条染色体中的大区域内。
10) 突变可以是自发的或诱导的。自发的突变的出现是由于DNA或染色体具有不稳定性。

【EX.4】 把下列句子翻译为中文

DNA被证明是遗传物质,Watson等人已破解了它的结构,但DNA是如何复制它的信息以及这些信息如何在表现型中表达出来尚未得到确认。Matthew Meselson和Franklin W. Stahl设计了一个试验来探明DNA复制的方式,结果认为有三种可能的复制模式。
1) 保留复制在复制时可能将以某种方式产生一条全新的DNA链。
2) 半保留复制产生两个DNA分子,每一个分子的一半是亲本DNA和另一半是一条全新的互补链。换句话说,新的DNA可能包括一条新DNA链和一条DNA旧链。已存在的链将作为新链的互补模板。
3) 散乱复制包括复制时母键的分开和以某种方式重新组合分子,组合分子的每条链上都是老的和新的片段的混杂。

【EX.5】 根据课文内容,回答以下问题

1) There are 46 chromosomes in the human genome. They are 44 autosomes and 2 sex chromosomes.
2) A common abnormality is caused by nondisjunction, the failure of replicated chromosomes to segregate during Anaphase Ⅱ.
3) Nondisjunction of one or more sex chromosomes may cause sex-chromosome abnormalities.
4) Now more than 3500 human genetic traits are known.
5) Albinism is the lack of pigmentation in skin, hair, and eyes.
6) Phenylketonuria (PKU) is recessively inherited disorder whose sufferers lack the ability to synthesize an enzyme to convert the amino acid phenylalanine into tyrosine.
7) Tay-Sachs Disease is an autosomal recessive resulting in degeneration of the nervous system.

8) Sickle-cell anemia is an autosomal recessive.

9) Huntingtou's disease, also referred to as Woody Guthrie's disease, after the folk singer who died in the 1960s, is an autosomal dominant resulting in progressive destruction of brain cells.

10) Polydactly is the presence of a sixth digit.

Unit 4

【EX. 1】 根据课文内容，回答以下问题

1) Microevolution is the changes in gene frequency that occur within a population without producing a new species.

2) Macroevolution is the evolution at the level of species or higher.

3) Microevolution can occur very quickly; indeed, it is probably always occurring while macroevolution occurs over much longer periods and is seldom observed within the human life span.

4) Natural selection then leads to the evolution of antibiotic-resistant strains.

5) This idea that the pace of evolution is not always slow and constant is referred to as punctuated equilibrium.

6) Chromosomal aberrations can introduce large changes in genes and the sequences that regulate them.

7) People think that a large comet impact triggered the extinction of the dinosaurs.

8) Classification now aims to group species according to their evolutionary history.

9) About 99 percent of the sequence of bases in the deoxyribonucleic acid (DNA) of chimpanzees and humans is identical.

10) The conclusion from anatomy is that chimpanzees and humans evolved from the same ancestor only a few million years ago.

【EX. 2】 根据下面的英文解释，写出相应的英文词汇

microevolution; macroevolution; species; evolution; aberration; anatomy; resistance; population; catastrophe; phylogeny; confirm

【EX. 3】 把下列句子翻译为中文

1) 生物进化是有机体随着时间的流逝而发生的群体遗传性质的改变。进化的出现一定是群体中个体之间存在遗传的差异。

2) 当具有一定性状的个体（如大个体）比其他个体留下更多的后代时，群体就发生进化。留下较多后代的个体的这种遗传性状在下一代中变得更为普遍。

3）适应性是一种有机体改善其在环境中的表现的特性。适应性是自然选择的产物，自然选择是具有特定遗传性状的个体由于这些性状存活下来，并以一种比其他个体较高的速度繁殖的过程。

4）地球上生命的多样性是物种不断分裂成两个或更多物种的结果。

5）当一个物种分裂成两个物种时，由于这两个物种是从同一祖先进化而来的，它们拥有许多共同的特征。

6）进化已经发生的证据是无法否认的。一个强有力的证据来自化石记录，那些化石记录为生物学家提供了推想地球生命演变的历史材料，并且展示了物种的演化的过程。

7）自然种群提供了进化演变的明确证据。

8）当检测基因或蛋白质差异性时，根据化石记录被认为是亲缘关系近的物种之间可能比亲缘关系远的物种之间具有更多的相似性。

9）几个间接的证据证明大进化已经发生，这一过程（包括同源结构、发育模式、退化结构、相似进化方式和分布方式）发生了连续变化。

10）尽管其反对意见普遍缺乏科学价值，达尔文的进化论还是在公众中受到争议。

【EX. 4】 把下列短文翻译为中文

只要性状至少部分可以遗传，性状就可以受自然选择而进化。然而，个体也可以经文化进化获得新的性状，文化进化是指从其他个体学习来的。文化进化在人类得到高度发展，人类的语言和非凡的学习能力使创新能够传播并被迅速地接受。但是经文化进化获得性状的惟一要求是个体必须具有学习能力。如鸟类模仿其他个体唱歌，导致歌的"语调"进化。猿的许多行为是通过学习传达的。在一个研究中，调查者比较了婆罗洲岛的四群猩猩和苏门答腊岛的两群猩猩。调查者证实24种行为只属于一个群体。这些行为与这些群体生活的环境差异没有关系。这些行为中的10个是特有的采食技巧，包括工具的使用。6个是社会信号的不同形式，如发出尖叫声。这样，一些个体模仿其他个体的行为使猩猩群文化特性得以发展。

【EX. 5】 根据课文内容，回答以下问题

1) Modern biology is based on several unifying themes, such as the cell theory, genetics and inheritance, Francis Crick's central dogma of information flow, and Darwin and Wallace's theory of evolution by natural selection.

2) Charles Darwin studied divinity and medicine.

3) Adaptation, Variation, Over-reproduction and survival of the fittest.

4) In 1836.

5) He eventually settled on four main points of a radical new hypothesis:

　　A) Adaptation: all organisms adapt to their environments.

　　B) Variation: all organisms are variable in their traits.

　　C) Over-reproduction: all organisms tend to reproduce beyond their environment's capacity to support them (this is based on the work of Thomas Malthus, who studied how populations of organisms tended to grow geometrically until they encountered a limit on their population size).

D) Natural selection: Since not all organisms are equally well adapted to their environment, some will survive and reproduce better than others.

6) In 1858, Darwin received a letter from Wallace, in which Darwin's as-yet-unpublished theory of evolution and adaptation was precisely detailed.

7) Evolutionary theory and the cell theory provide us with a basis for the interrelation of all living things.

8) The five kingdoms of living organisms are: Monera, Protista, Fungi, Plantae and Animalia.

Unit 5

【EX. 1】 根据课文内容,回答以下问题

1) A plant has two organ systens. They are the shoot system and the root system.

2) The shoot system includes the organs such as leaves, buds, stems, flowers (if the plant has any), and fruits (if the plant has any). The root system includes those parts of the plant below ground, such as the roots, tubers and rhizomes.

3) Plants have only three tissue types. They are dermal, ground and vascular.

4) A meristem may be defined as a region of localized mitosis.

5) A cambium is a lateral meristem that produces (usually) secondaty growth.

6) Collenchyma cells are characterized by thickenings of the wall, they are alive at maturity. They tend to occur as part of vascular bundles or on the corners of angular stems. In many prepared slides they stain red. Sclerenchyma cells often occur as bundle cap fibers. Sclerenchyma cells are characterized by thickenings in their secondary walls. They are dead at maturity.

7) Xylem is a term applied to woody (lignin-impregnated) walls of certain cells of plants.

8) Tracheids are long and tapered, with angled end plates that connect cell to cell. Vessel elements are shorter, much wider, and lack end plates. They occur only in angiosperms, the most recently evolved large group of plants.

9) The function of the epidermal tissue is to prevent water loss and acts as a barrier to fungi and other invaders.

10) Guard cells facilitate gas exchange between the inner parts of leaves, stems and fruits, plants have a series of openings known as stomata.

【EX. 2】 根据下面的英文解释,写出相应的英文词汇
xylem; parenchyma; phloem; tracheid; rhizome; tuber; sclerenchyma; epidermis; angiosperm; herbaceous

【EX. 3】 把下列句子翻译为中文
1) 分生组织是正处于分裂的组织,胚性组织负责初级和次级生长。
2) 植物基本的生活周期即世代交替可以归纳为:受精卵-孢子体-减数分裂-孢子-配子体-配子(卵子和精子)-配子结合-受精卵。这样的一种周期是所有植物的特征。
3) 植物依靠分生组织的分裂进行生长。初级生长是位于植物顶尖的顶端分生组织细胞分裂的结果,它使植株变长。
4) 叶子的生长是有限的,就像花一样;而茎和根的生长是无限的。在有限生长中,分生细胞最终会停止分裂;而在无限生长中,它们则保持无限分裂的能力。
5) 木质部将水分和溶解的矿物质从根向茎和叶传导。韧皮部将有机物质从植物的一个部位运输到另一个部位。
6) 运输系统、外部屏障和侧根系统是随着初级根的成熟发育而来的。
7) 有些植物有变态根,它具有进行光合作用、收集氧气、寄生于其他植物、贮藏食物或水,或者支撑茎的作用。
8) 被子植物的雄性和雌性结构常常出现在同一朵花中。这些繁殖结构不是成年植株的永久的一部分,在发育的早期也不出现。
9) 蜜蜂是最常见且独特的花粉传播者。昆虫常常被花的气味所吸引。鸟传播花粉的花是无气味的、红色的,这些花的芳香气味使昆虫不易接近。
10) 在双受精中,被子植物利用两个精细胞。一个细胞与卵细胞受精,而另一个细胞形成胚乳,胚乳为胚胎提供营养。

【EX. 4】 把下列短文翻译为中文
　　植物营养的主要来源是利用太阳能将大气中的二氧化碳转变为单糖。二氧化碳通过气孔进入植株。氧气是光合作用的产物,同样通过气孔排出植株,成为大气的成分。从糖分子的化学键中释放能量用于细胞呼吸作用,支持植物的生长和维持。然而,二氧化碳和光能不能满足植物合成所有分子的需要。植物还需要大量无机营养素。这些无机营养素中,植物需要量相对大的为常量营养素,需要量少的为微量营养素。有9种常量营养素:碳、氢和氧——这三种元素形成有机化合物,还有氮(合成氨基酸所必需)、钾、钙、磷、镁(叶绿素分子的中心)和硫。这些营养素中每一个的量,如碳,在健康植物的干物质重量中都远大于1%。铁、氯、铜、锰、锌、钼和硼这7种微量营养元素在多数植物中含量介于百万分之一到万分之几。

【EX. 5】 根据课文内容,回答以下问题
1) Flowers are collections of reproductive and sterile tissue arranged in a tight whorled

array having very short internodes.

2) Sterile parts of flowers are the sepals and petals. When these are similar in size and shape, they are termed tepals. Reproductive parts of the flower are the stamen and carpel.

3) The individual units of the androecium are the stamens. They consist of a filament which supports the anther.

4) The anther contains four microsporangia within which microspores (pollen) are produced by meiosis.

5) The gynoecium consists of the stigma, style and ovary.

6) The stigma functions as a receptive surface on which pollen lands and germinates its pollen tube.

7) The male gametophyte develops inside the pollen grain. The female gametophyte develops inside the ovule.

8) One sperm cell fuses with the egg, producing the zygote which will later develop into the next-generation sporophyte. The second sperm fuses with the two polar bodies located in the center of the sac, producing the nutritive triploid endosperm tissue that will provide energy for the embryo's growth and development.

9) Fruits may be fleshy, hard, multiple or single.

10) Runners are shoots running along or over the surface of the ground that will sprout a plantlet, which upon settling to the ground develop into a new independent plant.

Unit 6

【EX. 1】 根据课文内容,回答以下问题

1) Four types of tissues are discussed in the text. They are epithelial tissue, connective tissue, muscle tissue and nervous tissue.

2) The functions of epithelial tissue are lining, protecting and forming glands.

3) There are three types of epithelium. They are squamous epithelium, cuboidal epithelium and columnar epithelium.

4) The functions of epithelial cells are moving materials in, out, or around the body, protecting the internal environment against the external environment and secrete a product.

5) Connective tissue serves many purposes in the body: binding, supporting, protecting, forming blood, storing fats and filling space.

6) The two types of connective tissue are Loose Connective Tissue (LCT) and Fibrous Connective Tissue (FCT).

7) Cartilage forms the embryonic skeleton of vertebrates and the adult skeleton of

sharks and rays.

8) Muscle tissue facilitates movement of the animal by contraction of individual muscle cells.
9) Three types of muscle fibers occur in animals. They are skeletal, smooth and cardiac.
10) Nervous tissue functions in the integration of stimulus and control of response to that stimulus.

【EX. 2】 根据下面的英文解释,写出相应的英文词汇
epithelium; intestine; fibroblast; secrete; cartilage; axon; dendrite; platelet; neuron

【EX. 3】 把下列句子翻译为中文
1) 人类与其他动物的身体含有一个被横膈膜分成胸腔和腹腔的空腔。体细胞组成组织,组织组成器官,器官组成系统。
2) 上皮组织包括覆盖所有体表和腺体的膜。
3) 结缔组织是以其细胞之间基质中有大量的细胞外物质为特征。结缔组织可以是疏松的或者是致密的。
4) 骨骼肌能够使脊椎动物的身体运动。心肌使心脏产生动力,而平滑肌提供多种内脏功能。
5) 神经细胞有不同类型,但是所有的神经细胞都有接收、产生和传导电信号的特点。
6) 脊椎动物身体由具有不同特殊功能的细胞、组织、器官和器官系统组成。
7) 成年脊椎动物身体的 4 种基本组织——上皮组织、结缔组织、肌肉组织和神经组织——源自三个胚胎层。
8) 平滑肌是由纺锤状细胞组成的,分布于体内器官和血管壁中。
9) 神经胶质是具有不同功能的支持细胞,这些功能包括绝缘神经轴突以加速电刺激。
10) 骨骼肌和心肌都是条纹状的;然而骨骼肌受自发控制,而心肌是随意肌。

【EX. 4】 把下列短文翻译为中文
多数动物都有一个"管状消化管"的机体构造。这种机体构造有两个开口,其中一个开口是为食物进入身体准备的(口腔),而另一个开口则为废物排出体外准备的(肛门)。以"管状消化管"构造的动物对食物的消化和吸收率比具有"囊状"机体构造的动物高 10%。这种管状消化管的机体构造使消化道功能专门化。而"囊状"机体构造动物仅仅只有一个开口,用于食物的摄取与废物的排出。"囊状"机体构造的动物没有专门化组织或器官的发育。

许多(并不是所有)动物,在胚胎发育期有三层组织:内胚层、中胚层和外胚层。有些动物,最显著的是海绵,缺乏这些胚层组织。腔肠动物刺胞亚门(珊瑚和水母)仅仅只有其中的两层。扁形虫、线虫等都有三层组织。人类是三胚层的。

【EX. 5】 根据课文内容,回答以下问题
1) They depend directly or indirectly on plants, photosynthetic protists (algae), or

autotrophic bacteria for nourishment.

2) The unicellular heterotrophic organisms were at one time regarded as simple animals. They are now considered to be members of the kingdom Protista.

3) Ninety-nine percent of animals are invertebrates.

4) The animal kingdom includes about 35 phyla, most of which occur in the sea.

5) Animal cells are distinct among multicellular organisms because they lack rigid cell walls and are usually quite flexible.

6) Because their cells are flexible and their nerve and muscle tissues evolve.

7) Animal eggs are much larger than the small flagellated sperm.

8) Two subkingdoms are generally recognized within the kingdom Animalia. They are Parazoa and Eumetazoa.

9) Parazoa are animals that for the most part lack a definite symmetry and possess neither tissues nor organs, mostly comprised of the sponges, phylum Porifera.

10) Eumetazoas are animals that have a definite shape and symmetry and, in most cases, tissues organized into organs and organ systems.

Unit 7

【EX. 1】 根据课文内容,回答以下问题

1) Ecosystems include both living and nonliving components.

2) The living components include habitats and niches occupied by organisms. Nonliving components include soil, water, light, inorganic nutrients and weather.

3) The term ecological niche refers to how an organism functions in an ecosystem.

4) Food webs, food chains and food pyramids are three ways of representing energy flow.

5) The first law states that energy is neither created nor destroyed, but instead changes from one form to another (potential to kinetic).

6) The second law mandates that when energy is transformed from one form to another, some usable energy is lost as heat.

7) The pyramid of biomass is calculated by multiplying the average weight for organisms times the number of organisms at each trophic level.

8) A food chain is a series of organisms each feeding on the one preceding it.

9) There are two types of food chain. They are decomposer and grazer.

10) Food webs are networks of feeding interactions among species.

【EX. 2】 根据下面的英文解释,写出相应的英文词汇

niche; herbivore; carnivore; habitat; biomass; insecticide; chemosynthetic; parasitism; omnivore; estuary

【EX. 3】 把下列句子翻译为中文

1）氮被生物所利用的过程几乎都通过细菌的新陈代谢实现，这些细菌包括一些自由生活的和其他的在豆类与其他植物根部共生的细菌。

2）磷酸盐是相对不溶解的，大多数的土壤里只存有少量的磷酸盐。由于量太少经常会限制植物的生长。

3）能量在生态系统流动过程中，在每一层次中都会有一定量的损失。

4）初级生产力是绿色植物、藻类和一些细菌光合作用的结果。次级生产力是异养生物产生的新的生物量。

5）食物链的每一环节都会有大量的能量损失，这限制了食物链的长度。通常，生产力高的食物链会更长一些。

6）由于自然界通过食物网相互连接，每一个营养级的物种都会相互影响，这种影响对食物网的影响可能是正面的也可能是负面的。

7）物种的繁荣会提高生态系统的生产力，空间异质性和气候的稳定性促进了物种的繁荣。

8）生物体利用一系列的生理的、形态学的和行为机制来适应环境的变化。随着时间的变化，物种就能进化以适应不同的环境。

9）这些存在于相似气候区域的生物群落，不管是在哪里发现的，都很相似。年度的均温和降水的变化可作为当地生物圈变化的良好的预报器。

10）海水的循环重新分配热量，对大陆的西部进行增温。目前洋流的干扰如厄尔尼诺现象可以对全球的气候产生极大的影响。

【EX. 4】 把下列短文翻译为中文

生态系统的组成和分界

生态系统的自然物质包括曾经活着的有机物，如来自于树木的木头、腐化的植物材料、动物产生的废物和死亡的有机体。生态系统的自然物质也包括无机物，如矿物质、氮、水、全部的山、平原、湖泊和河流。

生态系统的生物体和自然环境相互作用。空气、水和土壤哺育着生物并限制着生存的生物种类，例如：淡水湖为一些特定的鱼类和水生植物提供了栖息的场所，但是同样的湖泊可能会杀死那些适应了咸水河口生活的动植物。

就像环境会影响生物体一样，生物体也会影响它们所生存的环境。苔藓会撼裂石头，树会挡住阳光，改变土壤的酸度和湿度，并释放氧气到空气中。大象可能会为了吃树叶而把树连根拔起，海狸堵塞溪流、形成草地，兔子啃咬地面上的草。

生态系统实际上不是封闭的，一个生态系统的界限是模糊的。如一个湖泊，会慢慢地演化成湿地，再演化成草地和灌木林。一股水流会从附近的森林带来营养物质和有机质，也会把当地的矿物质带离到其他的生态系统中。甚至大型的生态系统也会与其他的生态系统相互作用。种子会被风从一个地方吹到另一个地方,动物会迁移,水流和空气把有机物以及它们的产物和残留物从一个生态系统带到另一个生态系统。

所有的生态系统共同构成整个生物圈,它包括地球上的所有生物和它们所生活的自然环境。生物圈与其他的生态系统不同,它有明显的界限。生物圈覆盖了整个地球表面,从地底开始一直向上延伸,一直到达大气层。

【EX. 5】 根据课文内容,回答以下问题

1) The tundra and desert biomes.

2) Tropical rain forests occur in regions near the equator. The climate is always warm (between 20℃ and 25℃) with plenty of rainfall (at least 190 cm/year).

3) The temperate forest biome occurs in south of the taiga in eastern North America, eastern Asia, and much of Europe. The rainfall there is abundant (30-80 inches/year; 75-200 cm).

4) The shrubland biome is dominated by shrubs with small but thick evergreen leaves that are often coated with a thick, waxy cuticle, and with thick underground stems that survive the dry summers and frequent fires. Shrublands occur in parts of South America, western Australia, central Chile and around the Mediterranean Sea.

5) Grasslands occur in temperate and tropical areas with reduced rainfall (10-30 inches per year) or prolonged dry seasons. Grasslands occur in the Americas, Africa, Asia and Australia. Soils in this region are deep and rich and are excellent for agriculture.

6) Deserts are characterized by dry conditions (usually less than 10 inches per year; 25 cm) and a wide temperature range.

7) Winters are cold and short, while summers tend to be cool.

8) The taiga is noted for its great stands of spruce, fir, hemlock and pine.

9) The tundra covers the northernmost regions of North America and Eurasia, about 20% of the Earth's land area. The rainfall this biome receives is about 20 cm (8-10 inches) annually.

10) Aquatic communities are classified into two communities. They are freshwater (inland) communities and marine (saltwater or oceanic) communities.

Unit 8

【EX. 1】 根据课文内容,回答以下问题

1) At a low level of resolution, we can determine the amino acid composition of the protein.

2) We can determine the amino acid composition of the protein by hydrolyzing the protein in 6 N HCl, 100°C, under vacuum for various time intervals.

3) The N-terminus of the protein can be determined by reacting the protein with fluoro-

dinitrobenzene (FDNB) or dansyl chloride, which reacts with any free amine in the protein, including the epsilon amino group of lysine.

4) The amino group of the protein is linked to the aromatic ring of the DNB through an amine and to the dansyl group by a sulfonamide.

5) The C-terminal amino acid can be determined by addition of carboxypeptidases, enzymes which cleave amino acids from the C-terminal.

6) Amino acids with specific functional groups can be determined by chemical reactions with specific modifying groups.

7) In one, the protein is sequenced; in the other, the DNA encoding the protein is sequenced, from which the amino acid sequence can be derived.

8) The maximal length of the peptide which can be sequenced is about 50 amino acids.

9) Trypsin cleaves proteins within a chain after Lys and Arg.

10) Chymotrypsin cleaves after aromatic amino acids, like Trp, Tyr and Phe.

【EX. 2】 根据下面的英文解释,写出相应的英文词汇
resolution; hydrolysis; elute; quantitate; derivative; interval; disulfide; peptide; sulfonamide; cyclization

【EX. 3】 把下列句子翻译为中文
1) 物体由许多原子组成,每一个原子包括带正电的质子和中子的核,以及围绕在它们周围的带负电的电子,自然界有许多的元素,但是其中只有一部分组成生物系统。

2) 一种元素的核素不同之处在于它们的中子数的不同。有些核素具有放射性,衰变时放出射线。

3) 当两个核子共享一对或多对电子时,就会形成牢固的共价键,共价键具有空间方向性,这样就给予分子三维构象。

4) 非极性分子与极性分子(包括水)相互作用很弱。非极性分子与其他分子由很弱的键结合在一起,这个弱键称为范德华力。

5) 功能基团组成了大分子的一部分,它具有特殊的化学特性。功能基团结合的化学方式有助于我们理解含有这些功能基团的分子的特性。

6) 大分子有特异的三维形状,这种三维形状取决于其他单体的结构、特点和序列。

7) 蛋白质功能包括支持、保护、催化、转运、防御、调控和运动。蛋白质功能的行使有时候需要附着辅基。

8) 蛋白质中共有20种氨基酸。每种氨基酸由一个氨基、一个羧基、一个氢基和一个与碳原子相连的侧链组成。

9) 氨基酸通过缩合反应以共价键的方式连成多肽链,缩合反应发生在羧基和氨基之间。

10) 尽管脂类能形成巨大的结构,但这些聚合体不是化学大分子,因为其基本单位不是通过共价键连在一起的。

练习答案

【EX.4】 把下列短文翻译为中文

氨基酸的活化

氨基酸活化过程分为两个步骤,这个过程受氨酰-tRNA 合成酶催化。每一个 tRNA 和它所携带的氨基酸都可被特异的氨酰-tRNA 合成酶所识别,这就意味着至少有 20 种不同的氨酰-tRNA 合成酶存在。而且由于原核生物与真核生物的起始子 met-tRNA 与非起始子 met-tRNA 明显不同,因此,氨酰-tRNA 合成酶实际上至少有 21 种。

氨基酸活化作用以 ATP 作为能量来源,并且是在氨酰-tRNA 合成酶催化的两个步骤反应中发生活化作用。首先,酶把氨基酸连在 ATP 的 α-磷酸盐上,同时伴随着焦磷酸盐的释出。这个过程被称为氨酰-腺苷酸转化的中间过程。第二步中,酶催化氨基酸使其转运到 tRNA 的 3'-末端腺苷酸(A)的核糖 2' 或 3'-OH 上,结合生成有活性的氨酰-tRNA。虽然这些反应是完全可逆的,但焦磷酸盐的双水解却促进正反应。

【EX.5】 根据课文内容,回答以下问题

1) They are the endoplasmic reticulum and the Golgi apparatus.
2) Ribosomes (large RNA-protein complexes) are the sites for protein synthesis.
3) The ER lumen contains high concentrations of molecular chaperones to assist protein folding.
4) The Golgi does not contain molecular chaperons since protein folding is complete when the proteins arrive.
5) When inhibitors of ER glycosylation are added to cells, protein misfolding and aggregation are observed.
6) After two glucose residues are removed by glucosidase Ⅰ and Ⅱ, the monoglucosylated protein binds to calnexin (CNX) and/or calreticulin (CRT).
7) If a glycoprotein has not folded completely, it is recognized by a glycoprotein glucosyltransferase, which adds a glucose to it.
8) We understand protein and nucleic acid structure and synthesis better.
9) Human and yeast glycoproteins synthesis produce the same mannose core in the ER.
10) Human proteins made in yeast contain many Man residues.

Unit 9

【EX.1】 根据课文内容,回答以下问题

1) The threats mentioned in the text organisms face are invasion and infection by bacteria, viruses, fungi and other foreign or disease-causing agents.
2) They are recognition, specificity and memory.
3) Molecules that start immune responses are called antigens.

4) The body don't usually start an immune response against its own antigens because cells that recognize self-antigens are deleted or inactivated.

5) After an antigen is cleared from the body, immunological memory allows an antigen to be recognized and removed more quickly if encountered again.

6) APCs include macrophages, dendritic cells and B cells.

7) Cytokines are made by all immune cells and they control the immune response.

8) It refers to that antibodies travel through the body's fluids and attach to antigens, targeting them for destruction by nonspecific defenses.

9) The invader is usually removed before it has a chance to cause disease because some of the cloned TC cells and B cells produced during a primary immune response develop into memory cells.

10) HIV infection causes AIDS by attacking TH cells.

【EX. 2】 根据下面的英文解释，写出相应的英文词汇
immune; cytotoxic; immunodeficiency; spleen; inactivate; invade, cytokine; chemotherapy; venom; allergy

【EX. 3】 把下列句子翻译为中文
1) 机体的表面防御系统包括皮肤、消化道和呼吸道的黏膜层，它们能在许多微生物入侵器官之前将其清除。
2) 炎症反应通过增加感染位置的血流量，然后升高体温抑制细菌生长来抗感染。
3) 免疫反应能在感染部位激活防御细胞。
4) 非特异性免疫防御系统包括身体的屏障如皮肤、噬菌细胞、杀伤细胞和补体蛋白。
5) 动物的免疫系统是从非脊椎动物严格的非特异性免疫进化到哺乳动物的双重防御系统。
6) 皮肤的防御不仅仅体现在它是一道护卫机体的致密屏障，而且以其表面的化学分泌物进一步加强防御能力。
7) 身体上和解剖学的屏障可防止病原体侵入，它们是机体抗感染的第一道防线。
8) 先天性免疫不会对任何一种病原体产生特异性，它构成机体防御的第一道防线，它包括解剖屏障、生理屏障、内吞作用、噬菌作用和炎症反应等。
9) 获得性免疫反应有四种免疫特性：特异性、多样性、记忆性和自身/非自身识别。
10) 免疫是机体保护自身免受外来机体或物体（抗原）侵害的一种机制。脊椎动物有两种免疫系统：先天性免疫和获得性免疫。

【EX. 4】 把下列短文翻译为中文
　　获得性免疫对于刺激的反应具有高度的特异性，也就是有显著的"记忆性"。一般情况下，接触抗原以后，免疫系统会在5~6天内对这种抗原产生获得性免疫反应。以后如果再遇到同样的抗原，会引起记忆性的反应：对于同种抗原的第二次侵入，将会有更快的

免疫反应,而且更为强烈,还能更为有效地抑制和清除病原体。获得性免疫的主要组成包括淋巴细胞、抗体及其产生的其他分子产物。

由于获得性免疫反应过程需要一定的时间,因此当宿主与病原体接触之后的最先一段关键时期内,先天免疫就起着第一道防线的作用。通常情况下,在获得性免疫激活之前,大部分的微生物都能很快被健康机体的先天免疫系统的防御机制清除。

【EX. 5】 根据课文内容,回答以下问题

1) When the surface defenses of the vertebrate body are occasionally breached, the body uses a host of nonspecific cellular and chemical devices to defend itself. We refer to this as the second line of defense.
2) These devices all have one property in common: they respond to any microbial infection without pausing to determine the invader's identity.
3) The lymphatic system is a central location for the collection and distribution of the cells of the immune system.
4) The most important of the vertebrate body's nonspecific defenses are white blood cells called leukocytes. They circulate through the body and attack invading microbes within tissues.
5) Macrophages are large, irregularly shaped cells that kill microbes by ingesting them through phagocytosis, much as an amoeba ingests a food particle.
6) Neutrophils are leukocytes that, like macrophages, ingest and kill bacteria by phagocytosis.
7) They kill cells of the body that have been infected with viruses.
8) The cellular defenses of vertebrates are enhanced by a very effective chemical defense called the complement system.
9) The proteins of the complement system can augment the effects of other body defenses. Some amplify the inflammatory response by stimulating histamine release; others attract phagocytes to the area of infection; and still others coat invading microbes, roughening the microbes' surfaces so that phagocytes may attach to them more readily.
10) There are three major categories of interferons. They are alpha, beta and gamma.

Unit 10

【EX. 1】 根据课文内容,回答以下问题
1) Stem cells are cells that grow and divide without limitation, and given the proper signals, can become any other type of cell.
2) As a human embryo grows, the early cells start dividing and forming different, specialized cells such as heart cells, bone cells and muscle cells.

3) If the cells are harvested from an early embryo (about 5-7 days after conception) and nudged in a particular direction in the laboratory, they can be directed to become a particular tissue or organ.

4) Degenerative diseases start with the slow breakdown of an organ and progress to organ failure.

5) Degenerative diseases include stroke, diabetes, liver and lung diseases, heart disease and Alzheimer's disease.

6) Stem cells could provide healthy tissue to replace those damaged by spinal cord injury or burns.

7) One problem with stem-cell research is that the embryos are destroyed when the stem cells are removed.

8) The federal government will fund research using leftover embryos from fertility treatments, but will not support research using embryos created solely for research purposes.

9) In vitro (Latin, meaning "in glass") fertilization procedures often result in the production of excess embryos because a large number of egg cells are harvested from a woman who wishes to become pregnant.

10) Recent studies published in peer-reviewed literature suggest that most adult tissues have stem cells, that these cells can be driven to become other cell types, and that they can be grown indefinitely in the laboratory.

【EX. 2】 根据下面的英文解释，写出相应的英文词汇
replica; pregnant; uterus; implant; stroke; diabetes; osteoarthritis; pancreas; dilemma; consent

【EX. 3】 把下列句子翻译为中文
1) 近来的试验证明了不同哺乳动物分化器官克隆的可能性，这为首次进行家畜的转基因克隆打开了一扇门。
2) 干细胞的移植可能会使我们能进行损坏或丢失组织的替换，为目前还无法治疗的许多疾病提供治疗。目前的研究工作主要集中在组织特异性干细胞，组织特异性干细胞不存在胚胎干细胞那样的伦理问题。
3) 目前的试验为基因变种动物的克隆开辟了一条路，也说明了人类的克隆是可行的。
4) 改变农作物食品的基因，以提高农作物产量和营养价值，并延长保存期。
5) 现在有一个关注点就是：转基因食品可能对环境和消费转基因食品的人产生负面的影响。
6) 针对优良的生产性状的动物克隆已经实行，将来的某一天也可能进行人类克隆，但是不清楚克隆出来的人是否健康。

7) 通常自然的人类克隆是通过自发产生一样的双胞胎而发生的。
8) 已经从成年牛、山羊、鼠、猫、猪、兔子和绵羊成功地克隆出具有优良性状的下一代个体。
9) 自从在1998年分离出胚胎干细胞以来,世界各地的实验室已经对采用干细胞来修复受到破坏或丢失的器官的可能性进行了探索。
10) 这些胚胎干细胞之所以带来了令人激动的希望,是因为它们能发育成任何一种组织,我们就有可能实现对受到损伤的心脏和脊髓的修复。

【EX. 4】 把下列短文翻译为中文

Wilmut成功地克隆了完全分化的绵羊细胞,这对于基因技术是一件里程碑的大事。尽管他的方案效率不高(277次试验只有一次成功),但是他排除了所有的疑虑,确立了自己的观点,即成体动物细胞克隆是可行的。在接下来的4年里,研究人员极大地提高了克隆的效率。利用Wilmut试验的关键要点来克隆处于静止期的细胞,他们已经回到曾最先被Briggs和King采用的核移植步骤,这种方法行得通。许多哺乳动物已被成功克隆,包括老鼠、猪和牛。

转基因克隆将会对医学和农业产生重大的影响。动物和人类的基因可被用来生产珍稀的激素。比如,最近转基因的绵羊分泌的奶中有一种特殊的蛋白质。这种蛋白质称为α-1抗胰蛋白酶(对减轻膀胱纤维化症状有帮助)。这种羊可能被克隆,这将可以大大降低这种昂贵药物的价格。

【EX. 5】 根据课文内容,回答以下问题

1) Gene therapy is an experimental disease treatment in which a gene is delivered to cells in the body.
2) No, it is not, but it may become so as researchers solve the many technical problems it presents.
3) Effective gene therapy requires delivering the gene to each cell in which it acts, integrating the gene with the thousands of others on the chromosomes and regulating the expression of the gene.
4) The reason given in the text is that viruses have been designed by evolution to deliver their own genes to our cells.
5) Because it can carry a very large gene and will infect most cell types.
6) Adenosine deaminase deficiency affects white blood cells and causes severe combined immune deficiency ("bubble boy" disease).
7) This can be treated by removing white blood cells, inserting the adenosine deaminase gene into them, and returning the cells to the bone marrow.
8) Cystic fibrosis presents a much bigger challenge because it affects the airways and pancreas.

9) Duchenne muscular dystrophy is an even bigger challenge since it affects all muscles, and muscles make up 45 percent of the body. The only realistic treatment option in this case is systemic delivery, which poses the added challenge of preventing delivery to non-muscle tissue.

10) Long-term expression requires that the gene join the host chromosome.